The Flowers And Gardens Of Madeira

By

Florence Du Cane

The Flowers And Gardens Of Madeira
by Florence Du Cane

ISBN: 978-93-61426-65-0

Published by

DOUBLE 9 BOOKS

2/13-B, Ansari Road
Daryaganj, New Delhi – 110002
info@double9books.com
www.double9books.com
Tel. 011-40042856

This book is under public domain

ABOUT THE AUTHOR

Ella Du Cane (1874–1943) was a British artist best known for her watercolors of landscapes and exotic locations. Ella Mary Du Cane was the third and youngest daughter of Sir Charles Du Cane, a politician and colonial administrator, and his wife, Georgiana Susan Copley. She was the great-granddaughter of artist John Singleton Copley via her mother. She was born in Hobart, Tasmania, during the final year of her father's five-year time as Governor of Tasmania, just before the family returned to their country house at Braxted Park, Essex. Charles was appointed Knight Commander of the Order of St Michael and St George in Essex, where he also served as Chairman of the Board of Customs. Her sister, Florence, was born in 1869. Ella Du Cane rose to artistic prominence in 1893, when she participated in an exhibition of the elite New Society of Painters in Water Colours. Queen Victoria took a special interest in Du Cane's work, purchasing 26 pieces between December 1895 and August 1898. With success came opportunities to travel. A 1902 exhibition of watercolor drawings from the West Indies was followed by a 1904 display of Japanese paintings. In 1905, A & C Black commissioned Du Cane to illustrate The Italian Lakes (1905), which was described by Richard Bagot. The corporation also utilized numerous of her Japan photographs in John Finnemore's book.

CONTENTS

CHAPTER I
INTRODUCTION

The very name of Madeira (or island of timber, as the word signifies) brings to the minds of most people a suggestion of luxuriant vegetation flourishing in a damp, enervating climate. Such, indeed, was my own mental picture of Madeira before my first visit to the island. I expected to find every garden with the aspect of a fernery, moisture dripping everywhere, and the hills clothed with the remains of the primeval forests. The latter might possibly still have existed had it not been for the zeal of the discoverers of the island in making use of their discovery from a utilitarian point of view, and cutting clearings for the cultivation of the rich and fertile land. In order to clear the ground of the forests, which we are told clothed the island to its very shores, the drastic measure of setting fire to it was resorted to: hence the destruction (as old historians assert that the fire raged for over two years) of all the forests on the south side of the island.

Some feeling of disappointment entered my mind when I first looked on the Bay of Funchal. As compared to the wooded appearance the north of the island presents, the south side, viewed from the sea, appears to have much less vegetation. Large stretches of pine woods, it is true, have been replanted, and though they are used for timber, and are felled before they attain any great size, regulations exist which oblige any person who cuts down a tree to plant another in its place. Though I should imagine it is more than doubtful whether this regulation is carried out to the letter, the plantations are replanted, or the stock of timber would otherwise soon become exhausted. The fact that the south side of any island is naturally the most suited for cultivation has also led to the destruction of the woods, and on approaching the island it is very soon seen that every available inch of ground is cultivated in some form or another. The cultivation may take the form of some cared-for garden, where trees, shrubs, and creepers from the tropics may be flourishing side by side with more familiar vegetation, or may merely be the little terraced patch of ground surrounding the humblest cottage, where the harvest of the crop—be it sugar-cane, *batata* (sweet potato), or yam—is eagerly looked forward to, in order to eke out the very slender means of its habitants.

The feelings of Edward Bowdick, as described in "Excursions to Madeira and Porto Santo in 1823," must often have been re-echoed by many a visitor who sees the island for the first time: "To those who have visited the tropics nothing can be more gratifying than to find the trees they have there dwelt on with so much pleasure, and which are decidedly the most beautiful part of the Creation; to be reminded of the vast solitudes, where vegetable nature seems to reign uncontrolled and untouched; to see the bright blue sky through the delicate pinnated leaves of the mimosa, whilst the wood strawberry at its feet recalls the still dearer recollection of home; to gather the fallen guavas with one hand and the blackberry with the other; to be able to choose between the apples and cherries of Europe (which are so much regretted) and the banana—it is this feeling which makes Madeira so delightful, independent of its beautiful scenery and the constancy and softness of its temperature."

A DRINKING FOUNTAIN

Any feeling of disappointment that the traveller may have experienced from his first cursory glance at the island must surely be quickly dispelled on landing, especially if this should be in the month of January, when, having left the snows and frosts of Europe behind, after travelling for four days he is basking in the almost perpetual sunshine of so-called winter in Madeira. Lovers of flowers—and to those I most recommend a visit to the island—will find fresh beauties even at every turn of the street: the gorgeous-coloured creepers seem to have taken possession everywhere. Hanging over every wall where their presence is permitted will come tumbling some great mass of creeper, be it the orange *Bignonia venustus*, whose clusters of surely the most brilliant orange-coloured flower that grows completely smother the foliage; or the scarlet, purple, or lilac bougainvillea, whose splendour will take one's breath away, with its dazzling mass of blossoms. The great white trumpets of the datura, combined possibly with the flaunting red pointsettia blossoms, will quickly show the fresh arrival the bewildering variety of the vegetation—so much so that I cannot fail again to sympathize with Mr. Bowdick, who, writing on the subject, says: "The enchanting landscape which presents itself flatters the botanist at the first view with a rich harvest, and not until he begins to work in earnest does he foresee the labours of his task. What can be more delightful than to see the banana and the violet on the same bank, and the *Melia adzerach*, with its dark shining leaves, raising its summit as high as that of its neighbour, the *Populus alba*? It is this very gratification which occasions the perplexity, at the same time that it confirms the opinion, that Madeira might be made the finest experimental garden in the world, and that an interchange of the plants of the tropical and temperate climates might be made successfully after they had been completely naturalized there."

Since the above was written (1823) no doubt much has been done in the way of naturalizing plants from other countries, chiefly by the English, who are the owners of most of the principal gardens in and around Funchal. Many a plant and bulb from the Cape has found a new home in Madeira, and has spread throughout the length and breadth of the island, straying from gardens until they have now become almost hedgerow flowers; while at a higher altitude than Funchal, plants from England and other parts of Europe have also found a new resting-place.

It is not only to lovers of flowers, who, should they become the happy possessors of a garden in Madeira, will find in it a never-ending source of enjoyment, but also to those who wish to explore the natural scenery of the island, that I heartily recommend a visit to Madeira. Probably no other island of its size has such grand and varied scenery. Being only some thirty-three miles long and fifteen across even at the widest part, most people look incredulous when told of the inaccessibleness of some of the more remote parts of the island, picturing to themselves the possibility of seeing the whole island in one or, at the outside, two days by means of the now ubiquitous motor-car. These impatient travellers had better stay away from Madeira, for their motor-cars will be of no use to them, the gradients of the roads being too steep for any but the most powerful of cars, even if the roads themselves were not paved with the most unlevel cobble-stones. To anyone who has leisure to spend in exploring the island, merely for the sake either of admiring its scenery, or making a collection of the many ferns which adorn every nook and cranny of the deep ravines, I can promise ample reward; always supposing that they are sufficiently good travellers not to consider comfortable hotel accommodation as being an essential part of their expedition. Away from Funchal no hotels exist in Madeira; but if it is the right season of the year, and a spell of fine weather is reasonably to be expected, tent-life must be resorted to, or the primitive accommodation afforded by the engineers' huts in various districts, or rooms in the most primitive of village inns.

Enthusiastic admirers of the scenery of Madeira have compared its grandeur to that of the Yosemite Valley in miniature: its mountain-peaks, it is true, only range from 4,000 to 6,000 feet, but the abruptness with which they rise gives an impression of enormous depth to the densely wooded ravines. In an article on Madeira written by Mr. Frazer in 1875 it will be seen that he also compared its scenery to some of the grandest mountain scenery in the world. Writing of an expedition to the north side of the island, he says: "The beauty of the scene culminated at the little hamlet of Cruzinhas, whence we looked into a labyrinth of dark precipitous ravines, formed by the gorges of the central group of mountains, whose peaks, fortunately unclouded for a time, resembled in their fantastic ruggedness those of the Dolomites; but their sides being densely wooded with the sparkling laurel, and the ravines themselves more tortuous, we, I need hardly say, reluctantly came to the conclusion that even the Dolomite gorges could not equal them. There was none of the splendid rock-colouring of the Dolomites, but for deep-wooded ravines of deep mysterious gloom, descending from pinnacled mountains, it is a great question whether the Tyrol must not yield to Madeira."

CHAPTER II
PORTUGUESE GARDENS

I have often been asked whether the Portuguese have any distinctive form of gardening, and in answer I can only say that, though there is no attempt to compete with the grand terraced gardens of Italy or France, or the prim conventionality of the gardens of the Dutch, still the little well-cared-for garden of the Portuguese has a great charm of its own. Here, in Madeira, their gardens are usually on a very small, almost diminutive, scale, according to our ideas of a garden. In the mother-country, where they probably surround more imposing houses, they may attain to a larger scale, but of that I know nothing.

The love of gardening, unfortunately, seems to be dying out among the Portuguese in Madeira, and many a garden which was formerly dear to its owner, each plant being tended with loving hands, has now fallen into ruin and decay. The little paths, neatly paved with small round cobble-stones of a pleasing brownish colour, have become overgrown and a prey to the worst pest in Madeira gardens, the coco grass, which is enough to break the heart of any gardener once it is allowed to get possession; its little green shoots seem to spring up in a single night, and the labour of yesterday has to be again the work of to-day if the neat, trim paths so necessary to any garden are to be kept free from the invader. Or the box hedges, which were formerly the pride of their owner, have lost their trimness and regularity from the lack of the shears at the necessary season, and the garden only suggests departed glories.

Luckily, a few of these gardens still remain in all their beauty, and the pleasure their owners display in showing them speaks for itself of their true love of gardening.

The plan of the garden is usually somewhat formal in design, and as a rule centres in a fountain or water-tank, which serves the double purpose of being an ornament to the garden and of supplying it with water. The entrance to the garden is certain to be through a corridor, with either square cement and plaster pillars, or merely stout wooden posts, which carry the vine or creeper-clad trellis. The beds are not each devoted to the cultivation of a separate flower, as would be the case in an English garden, but single well-grown specimens of different kinds of plants fill the beds. Begonias, in great variety, tall and short, with blossoms large and small, shading from white through every gradation of pink to deep scarlet, form a most important foundation for every Portuguese garden; as, from their prolonged season of blooming, some varieties seeming to be in perpetual bloom, they always provide a note of colour. Pelargoniums, allowed to grow into tall bushes, in due season make bright masses of colour, the velvety texture of their petals seeming to enhance the brilliancy of their colouring. Fuchsias in endless variety, salvias red and blue, mauve lantanas, scarlet bouvardias, and *Linum trigynum*, with its clear yellow blossoms, help to keep the little gardens gay through the winter months. The latter, though commonly called *Linum*, is a synonym of *Reinwardtia trigynum* and a native of the mountains of the East Indies.

Last, but by no means least in importance, come the sweet-smelling plants, essential to these little miniature gardens. *Olea fragrans*, or sweet olive, also called *Osmanthus fragrans*, must be given the palm, as surely its insignificant little greenish-white flower is the sweetest flower that grows, and fills the whole air with its delicious fragrance. *Diosma ericoides*, a well-named plant—from *dios*, divine, and *osme*, small—ought perhaps to have been given the first place, as it will never fail at every season of the year to bring fragrance to the garden. The tender green of its heath-like growth, when crushed, yields a strong aromatic scent, and no Portuguese garden is complete without its bushes of *Diosma*. If allowed to grow undisturbed, it will make shrubs of considerable size, and in the early spring is covered with little white starry flowers; but as it bears clipping kindly, it is especially dear to the heart of the Portuguese gardener, who will fashion arm-chairs, or tables, or neat round and square bushes, in the same way as the Dutch clip their yew-trees. Rosemary also ranks high in their affections, not only for its sweet-smelling properties, but also because it can be subjected to the same treatment. Sweet-scented verbenas are also favourites, and in spring the tiny white flower of the small creeping smilax suggests the presence of orange-groves by its almost overpowering scent.

Camellias, white and pink, single and double, are favourite flowers, but as a rule the shrubs are subjected to drastic treatment and cut back, so as to keep the plants within bounds and in proportion to the size of the garden. Here and there a leafless *Magnolia conspicua* adorns the garden with its cup-like blossoms in the early spring, and a few other shrubs are permitted within the precincts of the garden. *Franciscea*, with its shiny green leaves and starry blossoms, shading from the palest grey to deep lilac, according to the time each bloom has been fully developed, should have been included in the list of sweet-smelling plants, as it has an almost overpoweringly strong scent. The bottle-brush, *Melaleuca*, with its strange reddish blossoms, showing how aptly it has been named, and the pear-scented magnolia, with its insignificant little brownish blossoms, are all favourite shrubs.

Various bulbous plants seem to have made a home under the shelter of their taller-growing companions, and in February, freesias, which in this land of flowers seed themselves, spring up in every nook and cranny; also the unconsidered sparaxis, whose deep red and yellow striped flowers are hardly worthy of a place. But the bright orange tritonias and deep blue babianas are highly prized, and in May the red amaryllis adorn most of the gardens, in company with the rosy-white *Crinum powellei*. The delicate *Gladiolus colvillei*, known in England as the Bride and under various other fancy names, open their pale pink-and-white spikes of bloom early in May. A few plants of carnations are treasured, as they are not easy to grow. Rose-trees are given a place, many being such old-fashioned varieties that I could not find a name for them; while the walls of the garden may be clad with heliotrope, which seems to be in perpetual bloom, or *Plumbago capensis*, whose clear blue blossoms cover the plant in great profusion in late autumn and spring. In summer the yellow blossoms of the *Allamanda Schotti* appear, and later in the year the waxy-white *Stephanotis floribunda* and *Mandevilleas* will all in turn be an ornament to the garden, though in the winter months their glossy green foliage will have passed unnoticed.

AZALEAS IN A PORTUGUESE GARDEN

I consider that *Azalea indica* is the plant which is most valued by the Portuguese. In the cared-for garden it is given a most conspicuous place, either planted in the open ground in partial shade, or more frequently kept in pots, and tended with the greatest care. In February and March through many an open doorway a glimpse may be caught of a group of gay-coloured azaleas, even in little humble gardens which at other seasons of the year are flowerless. The whole horticultural energy of the owner of the little strip of garden has been centred in the loving care bestowed on his few treasured azaleas. A tiny plant, not more than a few inches in height, will be far more valued than its overgrown neighbour, if it should happen to be some new variety, possibly only bearing a few blossoms, but perfect in form, of immense size, single or semi-double, of a brilliant rose-red, clear pink, salmon colour, or pure white. The culture of azaleas does not seem to be peculiar to the natives of Madeira, as from Oporto come numerous sturdy little trees of all the most highly prized varieties. The effect of well-grown specimens in pots, arranged along the stone ledge of the garden corridor, or grouped round the stone or, more correctly speaking, plaster seat, which generally finds a place in all these gardens, is very pleasing, and well repays the care bestowed on the plants all through the heat of the summer months.

A corner of the garden must be devoted to fern-growing, without which no garden in Madeira is complete. In the gardens of the rich a little greenhouse, or *stufa* is considered necessary for their successful cultivation, but in many a shady, damp corner of a humble cottage garden have I seen splendid specimens of the commoner ferns grown without that most disfiguring element. Perfect shelter from wind and sun is, of course, necessary, and sometimes, where no other shelter is available, the dense shade of a spreading Madeira cedar-tree is made use of, and from its branches will hang fern-clad pots. Or a little arbour is formed of that most useful of shade-giving creepers, the native *Allegra campo*, or Happy Country. The plant is also sometimes called Alexandrian laurel, though for what reason it is hard to know, as it has no connection with the Laurel family, but is *Ruscus racemorus*. The plant throws up fresh shoots every winter, which in their early stages appear like giant asparagus, and grow and grow until sometimes they reach fifteen or twenty feet in length before the fresh pale green leaves develop. By the spring the young leaves have unfurled, and provide a canopy of delicate green through the summer. The growth of the previous year can either be cut away, or if retained, in late spring, little greenish-white flowers will appear on the underneath of the leaves. The plant is a native of Portugal, but may be found in a wild state in Madeira. It is also known under the name of *Danæ racemosus*. One of the Polypodiums, called by the Portuguese *Feto do metre*, or Fern by the yard, seems to be first favourite, and splendid specimens are to be seen, each frond measuring one to two yards in length. *Gymnogrammes*, or golden ferns, are also much prized, and the *Asparagus sprengerii* has during the last few years found many admirers, with its long sprays rivalling in length the *Feto do metro*. *Adiantums* and all the commoner ferns are given a place, according to the taste of their owners.

I cannot close this chapter without a few words on the subject of the neat devices made by the Portuguese out of canes or bamboo, for training plants. In some instances it may be overdone, and one cannot always admire rose-trees trained on to bamboo frames in the shape of fans, crosses, or even umbrellas; but the little arched fences as a support to lower-growing plants are used with very good effect. I have copied the idea in England with some success for training ivy-leaved geraniums in large pots or tubs, by planting four rather stout bamboos or canes, two feet or more in height, in the pots, then slipping four pieces of split cane into the hollow ends, and either forming four arches, by inserting each end of the split length into the hollow, or else a pagoda-like effect can be made by taking the split canes into the middle, and then slipping all four ends through a hollow piece of cane a couple of inches long. Side arches can be made in any number, according to the requirements of the plant or the fancy of the gardener, by making incisions in the stout bamboos at any distance from the ground, and inserting the ends of the split canes. Old carnation plants, or seedlings which bear many flower-stems, may be very successfully and neatly supported in this way.

AZALEAS, QUINTA ILHEOS

Another contrivance for the increase of their rose-trees struck me as original, and worth mentioning, and possibly imitating, by those who garden in a subtropical climate—this is their system of layering rose-branches. My idea of layering carnations, shrubs, or any other plants, had always been to cut the plant at a joint, and peg it firmly into the ground, covering with a few inches of fine soil; but the Madeira gardeners adopt a different system, anyway, with regard to their roses. The branch for layering is not chosen near the ground, but often at a height of from two to four feet. The chosen branch is passed through the hole at the bottom of a flower-pot, or a box with a good-sized hole in it answers the same purpose; the pot or box is then supported at the necessary height on a tripod of sticks or bamboos. The branch has an upward slit made in the ordinary way, and the pot is then filled with soil. In two or three months' time, I was assured, the branch would be well rooted and ready to be transplanted to its fresh quarters. It seemed a simple method of increasing rose-trees, which, as a rule, in climates like those of Madeira, flourish much better when grown on their own roots than grafted on to a foreign stock. The same system appears to answer admirably for the increase of shrubs and even trees, and is extensively adopted for creepers, especially bougainvilleas, which do not strike readily from cuttings; so it is no uncommon sight to see pots lodging among the branches of trees, with a layered branch ready to form a new tree.

CHAPTER III
VILLA GARDENS TO THE WEST OF FUNCHAL

The miniature gardens described in the previous chapter, which, as a rule, surround the more humble dwellings of the Portuguese, frequently only cover the small piece of ground at the back of the town house, which is either converted into the backyard and rubbish-heap, decorated with old tins and broken china, or converted into a little paradise of flowers, according to the temperament and taste of its owner. Apart from these are the larger gardens surrounding the villas, or *quintas*, on the outskirts of the town. Most of these gardens are owned by English residents, and to them Madeira owes the introduction of many floral treasures. The first impression of these gardens, taken from a general point of view, is that they are lacking in form, the idea conveyed being that the original owner of the garden made it without any definite plan in view. For that reason they invariably lack any sense of grandeur or repose. It is only fair to say, however, that the landscape gardener has had many difficulties to contend with. The natural slope of the ground is, as a rule, extremely steep, especially in gardens situated on the east side of the town. But the ground by no means necessarily falls away only in front of the house. It as often as not falls to one side as well, which makes terracing a very difficult and serious undertaking. To move earth by means of small baskets carried on men's backs is a sufficiently serious matter in the East, where coolies are employed at a very low rate of wages, and are accustomed to this method. But in Madeira, where wages are by no means low, this procedure, which is absolutely necessary, has an important financial aspect when laying out a garden. The result is to give the gardens the effect of having been added to bit by bit, and many of them are broken by slanting terraces without any particular meaning. In common with all foreign gardens, they lack the beauty of English turf, as the finer grasses will not withstand the heat and dryness of a Madeira summer. Natal grass, which grows from very small tubers, is the most common substitute for turf, as it is hardy and can be mown fairly close. Some of the finer American grasses have been found successful, especially for growing under large trees, which is most useful, as nothing is so unsatisfactory as the effect of trees growing out of would-be flower-beds. All the beauty of the trees is lost through the outline of the stems being confused by the surrounding plants, which in themselves are probably poor specimens, owing to the fact that they are constantly being starved through the goodness of the soil being absorbed by the roots of the trees.

Stone balustrades are unknown in Madeira, where cement or plaster has to take the place of stone. Simple designs can be carried out by this means, but, as a rule, a low wall, only about two or three feet in height, from which rise at intervals square pillars, originally intended to support the wooden cross-bars of the vine pergola, finishes the terrace and gives it a very characteristic effect. These pillars can be creeper-clad, and either stand alone or support a canopy of wistaria, bignonia, or some other gorgeous creeper.

Any defect in the scheme of the gardens is amply atoned for by the wealth of colour and abundance of flowers they contain, at almost all seasons of the year.

Some of the older gardens were laid out more as pleasure-grounds, and planted with specimen trees brought together from all parts of the New and Old world, and in these especially the lack of good turf is keenly felt. I am thinking of the gardens which surround the Hospicio, which was built in 1856 by the late Empress of Brazil, in memory of her daughter, the Princess Maria Amelia, who died in Madeira.

The garden is well cared for, and contains a good collection of trees and flowering shrubs. Near the entrance are some very fine *Ficus comosa* and two splendid *Jacarandas*, which, when they are laden with their blue blossoms, stand out splendidly against the dark evergreen trees; also a very large Coral-tree, whose grey leafless branches are adorned early in the year with scarlet blossoms. In the centre of the garden are two unusually fine specimens of *Duranta* trees, whose long hanging racemes of orange berries cause them to be much admired all through the winter and spring months, while in summer the branches are laden with their blue blossoms. Dragon-trees, frangipani-trees, judas-trees, camphor-trees, til and *Astrapea viscosa*, are all to be found here, and a large specimen of the gorgeous flame-coloured Flamboyant or Poinciana, may be easily recognized by its flat spreading branches, which shed their fern-like foliage before the blossoms appear. At all seasons of the year the garden affords a delightful pleasaunce for the inmates of the Hospital, and can never be entirely colourless, as the red dracenas and the bright crimson leaves of the acalypha, which are blotches of a lighter or darker colour, afford a welcome note of colour at all seasons of the year and a relief to the eternal green of the evergreen trees. The walls of the garden are clothed with bougainvilleas, wistarias, and other creepers, and the beds contain a variety of plants, such as clerodendrons, hibiscus, abutilons, begonias, azaleas, and roses. The grass edges to the beds give the garden a character of its own, and might well be copied in other Madeira gardens.

DATURA, QUINTA VIGIA

On the opposite side of the same road at the top of the Augustias Hill stands the Quinta Vigia; the name means a look-out place or watch-post, and no doubt the villa was so called because the grounds command a fine position, the terrace wall ending with a sheer descent 100 feet or more down to the sea. The garden has a fine view of the harbour, the Brazen Head, the distant islands of the Desertas, and the Loo Rock, which lies immediately below the cliff. The late Empress of Austria spent the winter of 1860-61 at this quinta, and since then the property has had various owners. Though the garden is now neglected, as the villa has been uninhabited for some years, the trees remain, and together with those belonging to the adjoining Quinta das Augustias on the one side, and those of the Quinta Pavao on the other, form one of the principal features of the town of Funchal. The day is probably fast approaching when the whole of this property will fall into the hands of an hotel company, but it is to be hoped that some effort will be made to save the trees. From far and near the splendid specimens of *Araucaria excelsa* form a very important feature in the landscape, as they have attained an immense size. I am told that Mr. William Copeland first introduced these Norfolk Island pines to Madeira, and planted those at Quinta Vigia. They seem to have taken kindly to their adopted country, though not, of course, attaining to the gigantic height of 150 feet, as they are said to do in their native land. The garden also contains a good specimen of *Araucaria braziliensis*. One of the largest cabbage palms in the island stands near the entrance, and the garden is rich in rare trees. Grevilleas, with deep orange bottle-brush-like flowers; schotia, with its deep crimson-red blossoms; magnolias; the deciduous *Taxodium distichum,* mango-trees, and hosts of others, adorn the grounds. Among the shrubs are pittosperums, with their leathery grey-green leaves and greenish-white sweet-scented blossoms; also francisceas and great quantities of *Euphorbia fulgens,* whose long wreaths of orange-scarlet flowers remain in beauty all through January and February. Here are to be seen pittangas, or *Eugenia braziliensis,* the myrtles of Brazil, with their small shiny foliage and little sweet-smelling white flowers, resembling the common myrtle, only borne on slender stalks; the ribbed orange-coloured fruit is not only very decorative to the shrub, but is valued as a great delicacy among the Portuguese. *Murraya exotica* has flowers closely resembling orange blossoms in form and fragrance, and appears to flower in spring and autumn. The verandah of the house is clothed in creepers, among which are *Allemanda schottii,* with its pure yellow blossoms, the deep crimson-flowered *Combritum; Thunbergia laurifolia,* with its lavender-coloured flowers; and *Rhyncospermum jasminoides,* whose tiny white starry flowers fill the whole air with their delicious fragrance late in April. Large bushes of the sweet-scented diosma and a small heath are a feature of the garden, while the great number of rose-trees are a legacy of one of its English owners, and in spite of the fact that they are now no longer carefully pruned, they flower in great profusion on immense bushes in December, and again in April.

A GROUP OF SENECIO

Near the entrance some large masses of purple and scarlet bougainvillea are to be seen, and by the middle of March the great buds of the immense and rampant-growing solandra are swelling, and in a few days the greenish-white trumpet-shaped flowers will have opened. The beauty of each individual blossom is short-lived: when newly opened it is of a greenish-white, which gradually turns to a deep cream colour, and then, alas! to a most unsightly brown. Unfortunately, the plant shares with thunbergia its ungraceful habit of retaining its blossoms in death, which mars the beauty of the freshly opened flowers. Large clumps of the yellow-flowered *Senecio grandifolia* are very effective when the great loose heads of blossom are at their best in February. The plant has fine foliage, and though many people despise it and regard it as a weed, on the outskirts of the garden or hanging over a wall it is certainly worthy of a place. Like its humble relation the common groundsel, it has an objectionable habit of scattering its fluffy seed to the four winds of heaven as soon as the plant is out of flower. This, to be sure, could be avoided by cutting off the old flower-heads as soon as they are over, but would be rather a Herculean task in gardens where it has spread into great beds. The plant is impatient of drought, and its foliage soon flags in the heat of the sun unless its roots are well supplied with moisture, and it will be discovered that its roots run far in the ground in search of it, which, combined with its practice of seeding itself in undesirable situations, makes it a dangerous plant to introduce unawares to a garden, as, once established, it is there for good.

Farther to the west of the town are the gardens of Quinta Magnolia, which cover an extensive area, largely increased by the present owner, until they now extend down the slope of the hill to the bed of *Ribeiro Secco*, or the Dry River. To those interested in the culture of palms these grounds will be of great interest, as the collection is a good one, and includes some very fine specimens, seen to great advantage, standing on slopes of the nearest approach to turf which the island can produce.

WEIGANDIA AND DAISIES

Some of the cabbage and date palms have attained an immense size, and are a great ornament to the landscape, and some fine groups of the curious screw pine, *Pandanus odoratissima*; it has peculiar flat leaves and an uncouth flower, which bears a strong resemblance to the body of a dead rabbit hanging from the plant! The grounds command fine views, and were laid out for the present owner by an English landscape gardener. There is a curious cave or grotto formed out of the natural rock, clothed with ferns and mosses, which no doubt remains cool and damp through the summer, and forms a welcome retreat from the fierce heat of the sun.

Close by are the grounds of Quinta Stanford, or Quinta Pitta, as it was originally called by its first owner. The gardens have been very much enlarged by their present owner; banana plantations have gradually vanished, and the grounds no longer present the cramped appearance from which they formerly suffered. New-comers to Madeira, as a rule, express great surprise that the gardens are not larger and generally only cover such a very small piece of ground. The value of the land for agricultural purposes—formerly for growing vines, then, possibly, for banana cultivation, and now for sugar-cane — is no doubt largely responsible for this, and also the great difficulty of acquiring a piece of ground of any considerable size in the neighbourhood of Funchal. In many cases even one acre may be owned by several different landlords, land being divided into incredibly small holdings.

CYPRESS AVENUE, QUINTA STANFORD

In this respect the owners of Quinta Stanford are to be envied, as the house stands well surrounded by its own ground, out of sight of the too common unsightly *fazenda* and its inevitable squalid cottages. From the terrace in front of the house the view is unrivalled, comprising a fine view of the sea and an unbroken view of the mountains behind the town of Funchal. It is easily seen that the garden is tended with unceasing care by its present owner, and near the entrance some judicious massing of shrubs and flowering trees has in a very few years well repaid the planter; some large clumps of weigandias, *Astrapea pendiflora*, and bushes of common white marguerite daisies of mammoth proportions give a broader effect than is usual in most Madeira gardens. To my mind, the very greatest praise should also be given to the owner for having planted an avenue of cypresses, almost the noblest and grandest of all trees, especially when seen under a southern sun, and their absence in the landscape of Madeira is keenly felt. The Portuguese see no beauty in them, and only connect them with death, for which reason they are scarcely ever seen except in cemeteries. From the astonishing growth which the young trees at Quinta Stanford have made in a few years, it is evident that the soil is very favourable for their culture, and it seems almost incredible that more owners of gardens, who must have seen what Italy owes to her cypresses, should not have planted them in

Madeira; but it is to be hoped that even now others may follow the excellent example set before them at Quinta Stanford.

The owner of the garden has much to tell of the successes and failures he has made, not only with imported plants, in the hopes of inducing them to find a new home in Madeira, but he journeyed far and wide to make a collection of the native ferns, of which there are a great quantity. Many of them, removed from the cool, damp air of their mountain homes, pined and died a lingering death in the air of Funchal, which was too hot and dry; and the atmosphere of a *stufa*, or greenhouse, is unsuited to the hardier ferns.

Some interesting experiments have also been made with rock-plants, in order to see whether it would be possible to induce any of our favourite Alpine plants to adopt a home in warmer climes; but I fear, though some may survive for a year or two, in the end they will grow steadily smaller, until they dwindle away and cease to exist. So I am afraid the making of a rock-garden in the sense which we in England regard a rock-garden— *i.e.*, an artificial arrangement of rocks, clothed with carpets and cushions of flowering Alpine plants—will never be possible in Madeira. Here the rock-garden must remain as Nature intended it to be—rocks and cliffs, interspersed with prickly-pear, agaves, cactus, some of the larger saxifrages, and such native plants as *Echium fastuosum*.

The gardens owned by the English suffer, as a rule, somewhat severely from the absence of their owners just at the season of the year when they require the closest care and attention, and this may possibly account for the failure to acclimatize many of these imported treasures. If they could be tended with loving hands, screened from the fiercest of the sun's rays, given exactly the amount of water they require, no doubt there would be many less failures; but the ignorant Portuguese gardener probably either starves the plant by entirely omitting to water it, especially if it is unlucky enough to be out of reach of the hose, or else he drowns it with too much water, until the ground surrounding it becomes a swamp: for the conditions suitable to a rock-plant would be as unknown to him as the conditions required by a bog-plant.

Some tree-ferns in a sheltered corner make a very good effect, and seem likely, from the strong growth they have made in a few years, to become very fine specimens.

On the terrace near the house are beds of begonias, roses, geraniums, heliotropes, sweet olives, and the garden flowers common to most Madeira gardens, while the walls are clad with a succession of creepers; so all through the winter months the garden remains a feast of colour.

Eighteen years ago the ground which is now the beautiful garden of the Palace Hotel was nothing but rocky, waste ground, bare of vegetation, except for the clumps of prickly-pear, agaves, and cacti which take possession of all the rocky ground along the shore. For situation the garden is unrivalled, and though the garden lacks the care and attention which naturally are bestowed on a private garden, the luxuriant growth, especially of the creepers, has converted the formerly waste ground into a beautiful jungle of flowers. The garden is devoid of any fine trees, except for the ficus trees, a few oaks, and a stray cypress or two which surround the Dépendence, which was formerly a private house; it stands at the very edge of the precipitous cliff, where the unceasing roar of the surf rings in one's ears as it dashes almost against its very walls. In front of the main building are some large cabbage palms, affording welcome shade and shelter, which have made astonishingly rapid growth, as only ten years ago they were merely items in flower-beds, and I little thought that on my second visit to the island, some seven years later, they would have become an important feature in the garden.

ALOES AND DAISY TREE

Early in December, when the whole island is fresh and green after the autumn rains, and presents more the aspect of spring than late autumn or even winter, the view from the garden is surprisingly beautiful. The cliffs have broad stretches of the brilliant red-flowered *Aloe arborescens*, with its large rosettes of glaucous grey-green leaves, which makes the plant always ornamental, even when it is not adorned with its hundreds of scarlet flower spikes. Some people say it was always indigenous to the island, and found its home in the Santa Luzia ravine. Whether this is really the case I feel doubtful, as Mr. Lowe, in his "Flora of Madeira," quotes it as one of the plants which has become naturalized, though probably originally introduced. Growing on the cliffs the flowers show to great advantage, standing out in sharp contrast to the deep blue sea below, but it is a great ornament wherever it grows, whether in clusters overhanging a wall where its rosettes of leaves overlap each other in thick tufts in endless succession till there seems no reason why they should ever stop, or clothing the rocky ground on the hillside among the pine-trees.

At the same season the *Franzeria artemesioides*, or daisy-trees, as they are commonly called, are in full beauty. The best method of treating these trees is to cut them back when they have done flowering, as the large clusters of daisy-like flowers appear on the long shoots of young wood. When their flowering season is over, they lose their large grey-green leaves, so it is lucky that the tree can be so treated, or the long bare branches would make them unsightly at other seasons. The hedges and bushes of *Plumbago capensis* attain to mammoth proportions when they can escape the attention of the gardener's ruthless shears, and are laden with their lovely soft blue blossoms in late November and December. Then comes a season of rest, though the plant is seldom entirely devoid of colour, and in early spring fresh shoots give promise of a wealth of blossom again in April and May.

Bougainvilleas have been planted with a lavish hand, but unluckily with no regard for colour. I sometimes wondered if the Portuguese gardeners are *all* colour-blind, as it is by no means uncommon to see a bright purple bougainvillea planted side by side with a scarlet one, and as likely as not, interlaced with a flaming orange bignonia, while the bright pink Charles Turner geranium grows happily below. In Madeira gardens colour runs riot, and I own that the prolonged flowering season of many of the creepers and shrubs makes the colour scheme more difficult than it is in our English gardens.

The great clumps of *Crinum powellei* are a remarkable feature of this garden, when late in April the great bulbs send up their spikes of either pure white flowers or white delicately flushed with pink. The flowers come in six to ten in an umbel, on stems three to five feet in height, and are very freely produced—large clumps sending up a dozen or more flower-heads at the same time. The bulb has long narrow green foliage, which is very ornamental. The flowers have a delicate but somewhat sickly scent; the plant is a native of Natal, and, like others of its compatriots, has taken kindly to the climate and soil of Madeira.

It would be impossible to enumerate all the host of other plants the garden contains—creepers, shrubs, flowering trees, besides roses, begonias, geraniums, heliotropes, in an almost endless list—while the cliffs have remained a natural rock-garden. In the clefts of the rocks giant agaves occasionally throw up their great flower-heads fifteen feet or more in height, and then the plant, as if exhausted by the supreme effort in the climax of its existence, dies; but it is quickly replaced by hundreds of others, as the seed of the monster flower has found fresh ground in every nook and cranny. Besides the agaves, clumps of prickly-pear, or *Opuntia tuna*, with its curious succulent growth clothed with poisonous thorns, some wild saxifrages and tufts of *Echium fastuosum*, known as Pride of Madeira, have all found a home.

This garden is the last one of any interest on the west side of the town, as beyond lie only a few modern villas in the worst possible taste, with no grounds worthy of the name of a garden; but almost opposite to the hotel in the grounds of Casa Branca for a few short weeks in the year the avenue of *Poinsettia pulcherrima* interspersed with date palms and clumps of strelitzias is worth seeing. The poinsettia blooms are almost the largest I have ever seen, measuring quite eighteen inches in diameter from point to point of the scarlet leaves. Like the daisy tree, the poinsettia flowers on the young wood, and throws out fresh branches six to ten feet long, which can be cut back in January, when the beauty of the blossoms is gone and the foliage becomes an unsightly yellow and at length drops altogether. When seen growing in all their luxuriant and garish splendour, it is difficult to remember that it is the same plant that one has seen in a weakly and attenuated form in our English stove-houses, with one poor little flower-head at the end of a single stem imperfectly clad with sickly foliage. Poinsettias seem to rejoice in rich soil, and they appear to revel in the liberal feeding of the adjoining banana plantations, which, no doubt, they deprive of a good deal of nourishment; but they well repay their owner, as in the glow of the western sun they provide a veritable feast of colour all through December.

POINSETTIA ON THE MOUNT ROAD

CHAPTER IV
VILLA GARDENS TO THE EAST OF FUNCHAL

On the east side of the town lie many quintas with good gardens, especially up the very steep Caminho do Monte, or Mount Road, as it is commonly called by the English. The road itself at some seasons of the year is converted into a veritable garden, as its high wall is so clothed with overhanging creepers which have strayed from the gardens behind, that it presents more the aspect of the terrace wall of a flower garden than that of one of the most frequented highroads of a town. At a height of between 500 and 600 feet, just below the level road which crosses it, which is known as the Levada da Santa Luzia, several villas seem to vie with each other as to which can contribute the greatest wealth of plants to decorate the walls. Possibly the best moment to see the road is in December, when the gorgeous mass of colour provided by the great shrubs of poinsettias hanging over the walls of the Quinta Santa Luzia is in all its splendour. Side by side with the scarlet blossoms come the great white trumpets of the daturas, hanging in horizontal rows. Below, the deep rose-coloured buds of the bougainvillea have not yet unfurled, so there is no jarring note in the scheme of colour, as the immense bank of plumbago, with its soft blue blossoms, harmonizes admirably. On the other side of the road, as if determined to continue the effect of the flaming red, is a great cluster of *Aloe arborescens*, with their spikes of red flowers—not, it is true, as brilliant in colouring as their opposite neighbours, the poinsettias, but very beautiful in themselves. These, with clumps of sweet-scented geraniums, echiums, and many other plants, clothe the walls of the garden of the Quinta da Levada. But the stream of gorgeous colour is not yet complete, as the *Bougainvillea spectabile*, with its brick-red blossoms, is already giving promise of glories to come in a very short time. This plant, which covers the corridor and hangs over an old garden well at Quinta Sant Andrea, is the finest specimen of its kind in the island. From the immense size of its stem, it is easily seen that the plant must be a great age, and for many years has borne its burden of blossoms and called forth the admiration of untold numbers of tourists through successive winters, as they make their noisy descent in the basket sledges or running cars from the mount.

THE SCARLET BOUGAINVILLEA

In a few weeks the road is turned into a golden road. The poinsettias and aloes will have shed their blossoms, and are soon forgotten, as the brilliant orange bignonia clothes many a wall and corridor, and in its turn attracts all attention. By April the wistaria takes its place, and the road becomes all mauve, as nowhere in the whole of Funchal are there so many beautiful wistarias collected together; all along this road they seem to have been planted with a lavish hand. Possibly the soil is especially suited to them in this district, as I have often heard owners of gardens in other parts of Funchal regret that they have never been able to establish this most beautiful of all creepers in their gardens.

It is small wonder that the sight of these flower-clad walls fills many a visitor to the island with a longing to see the gardens they enclose.

The palm must be given to the garden of Santa Luzia, as not only does it cover a much larger expanse of ground than any other, but the owner takes so much individual interest in almost every plant in the garden, that here, as it is always said flowers grow better for those who love them, everything seems to flourish and grow at its best. Like all good gardeners, she has not been deterred by the failure of a plant one season or the failure to import a new treasure at the first attempt, but has given hosts of plants a fair trial, often rewarded with success in the end, though naturally failing in some cases. Plants have been sent to her from all parts of the world, and the island owes many of its flowery treasures to this garden, which was originally their nursery and trial-ground. One of the most remarkable instances of this is *Streptosolen Jamesonii*, originally introduced to this garden, but which only succeeded the fourth time it was imported, and has now spread, until there is hardly a humble cottage garden in the whole of Funchal which is not decorated with its orange bushes in the winter months. The garden has been much enlarged of late years, and gradually terrace after terrace has been added to it, many of them forming a complete little garden in themselves. From the lie of the ground in a steep slope in two directions, and possibly from the fact that the garden has been added to gradually, it shares the difficulty I have described elsewhere, and had no very imposing scheme to start with.

WISTARIA, SANTA LUZIA

The entrance to the garden leads one to expect a wealth of flowers in the garden below, as a vision of pink begonias with a profusion of blossom, tall feathery bamboos, and long hanging ferns, greets the eye at the very door. On the terrace in front of the house stands one of the finest wistarias. It clothes the whole wall, makes a purple canopy to the corridor, climbs up the square pillars, and has even taken possession of the flagstaff, so in the early days of April the whole air is filled with its delicate bean-like scent. The beauty of its blossoms is short-lived, and possibly for this reason is all the more appreciated. A few short days and the heat of the sun will have taken all the colour out of its purple tassels, the leaves will begin to appear, and all its glory is departed. Some of the winter-flowering creepers last in beauty so long—for weeks or almost months—such as the bougainvilleas and *Bignonia venustus*, that if such a thing were possible, one becomes almost wearied of their beauty, and passes them by almost unnoticed. But with wistaria it is different: it must be noticed and appreciated at once or not at all, as the colour changes and fades with every passing hour.

Possibly April is the best month to visit this garden, though at no season is it without flowers, but March, April, and May are the best months of the year in all Madeira gardens. In some ways the autumn here seems as though it ought to be spring. Late in September or early in October the gardens go through the tidying up, pruning, and cutting back, which is generally done in our English gardens in early spring, and are made ready to reap the full benefit of the heavy autumn rains. Here during the summer everything has been left to grow as it will: the roses put forth long, rank, flowerless growth; the creepers grow out of all bounds; geraniums grow "leggy," with long leafless stems; the heliotrope has flowered itself to death, and must be cut back in order to make fresh growth for the coming season. The gardens by the end of the long, dry summer must present the aspect of an overgrown jungle, and according to the judicious or injudicious pruning in September and October will greatly depend the failure or success of the garden for the rest of the year. This also is the season for sowing seeds, and probably the best moment for starting newly imported treasures; it is most important that all these operations should be got through early in October, as by November it is soon evident that it is not really spring; the sap is not really rising, and through December, January, and February, it lies more or less stagnant and dormant, so unless seedlings and cuttings have made a good start before then, they will grow but little during those three months. The same will apply to plants which have been cut back; they should have made fresh shoots before the middle of November, or they will remain more or less bare and unsightly throughout the winter. By the time when most of the English owners return to their gardens in late November or early December, all traces of the necessary cutting should have vanished, and though the garden may not be gay with flowers, it should be full of promise of glories to come. But it seems hard to train a Portuguese gardener to get through his pruning at this season, and to have done with it for the time being, as, according to his ideas, pruning should be done apparently promiscuously, at any and every season of the year, and he is never happy without a pruning-knife in his hand, as often as not dealing death and destruction to a plant when it is in full beauty.

In the lower part of the garden a small pond, shaded by a weeping willow, whose parent was grown from a cutting brought from Longwood, provides a home for the white, pink, and blue water-lilies, which, with a large clump of papyrus, speedily remind one that one is in subtropical regions, where no breath of winter will ever reach the water sufficiently to bring death to the blue lilies which we in England know as pampered flowers, and can only grow by providing them with a warm bath, heated by artificial means.

On one of the terraces broad sheets of the mauve Virginian stock—with us an unconsidered little flower, but here, from the sheer wealth of its blossoms, providing a mass of colour—lead to a little Iris garden. Only the white *Iris Florentina* and a deep purple *Iris Germanica* really seem to flourish, so the beds are filled with these two kinds only. *Iris Pallida* and many of the other beautiful varieties of *Iris Germanica* have refused to make a home here, so the two kinds only have been retained, and for a few weeks in late December and early January the little garden is all purple and white. The purple weigandia flowers and the white of the Porto Santo daisy-trees help to carry out the colour scheme. The walls of the little garden are clad with the old Fortune's yellow roses, called by some Beauty of Glazenwood, and it is certainly one of the roses which thrive best in Madeira, bearing its burden of yellow and pink-tipped blossoms in the spring. On the corridor above a host of creepers flourish, but the blossoms of the Burmese rose were new to me. Its large single blooms open a delicate lemon colour, which gradually turns to white, and its shiny foliage is also very ornamental; but I fear its constitution will never stand the cold of our English winters, or even if it survived the cold, the warmth of our summers would not be sufficient to ripen the wood enough to make it flower. I believe it to be the same rose which has been grown with some success on the Riviera under the name of *Rosa grandiflora*. Near by is its fellow-countryman, the Burmese honeysuckle, suggesting a monster form of French honeysuckle; the foliage of its long twining branches closely resembles it, only on a very large scale, and the white trumpets of its blossoms, instead of being one or one and a half inches long, are from four to five inches in length. The heavy scent is almost overpowering, coming at a season of the year when the air seems to bring out the scent of the flowers to such an extent that they become almost offensive.

The garden is so full of interesting trees and shrubs that it would be a hopeless and never-ending task to attempt to enumerate them all, but the curious trunk and roots of all that remains of a formerly grand specimen of a Bella Sombra, or *Phytolacca dioica,* attract the attention of all new-comers. From the uncouth root have sprung numerous fresh branches, but they can never make a fine tree like their original parent. As a foliage plant *Monstera deliciosa,* a native of Mexico, makes a fine group where it can be allowed sufficient space to throw out its long aerial roots, by which it will firmly attach itself to a wall or bank. It must have been these strange roots which gained for it the first part of its name, as its deeply perforated dark green leathery leaves are no monsters, and I imagine it owes the second part to its fruit, which I have seen described as being "succulent, with a luscious pine-apple flavour."

There is a very fine specimen of *Bombax,* or silk cotton tree, which has a peculiar growth, and in June is covered with fluffy white blossoms.

ROSES, SANTA LUZIA

At again a lower level on yet another terrace is a little sunk garden, which seems to provide a never-ending wealth of colour and blossom. Between its box-edged beds run narrow walks, paved with flag-stones, a welcome relief to the usual paving with little round cobble-stones, and certainly pleasanter to walk upon, and in spring, when flowers spring up in every direction, many a little treasure appears between the stones. One I remember I could never regard as a weed, though many people seemed merely to look upon it as such, was *Anamotheca cruenta,* a tiny little bulb which bears very brilliant salmon-pink blossoms in clusters of five or six, each with a deep crimson mark in it. It is a native of the Cape, from where it was no doubt originally imported, and seems to sow itself freely. The borders are devoted to large clumps of such plants as eupatoriums, salvias, euphorbias, pelargoniums, albizzias, justicias, begonias, crinums, and imantophyllums, while in the centre of the garden rose-beds carpeted with freesias, and beds of the dark purple heliotrope, pink begonias, and lilac stocks, provide good masses of colour. Over the wall at one end of the garden, which is the boundary wall of the garden proper, hang great bushes of poinsettias, daturas, and large clumps of echiums, and on the top of the low wall on the other side, large pots of azaleas, diosmas, begonias, and ivy-leaf geraniums stand with very good effect.

Yet another of these little terrace gardens has been devoted entirely to the culture of blue and white flowers, which is a pretty idea, though true blue flowers are scarce. Blue salvias and solanums, justicias and linums are a good foundation for the garden, which, again, has paved walks, into whose cracks innumerable treasures have sown themselves. Freesias, violets, which, though not true blue, are too sweet to be ruthlessly weeded out, and forget-me-nots seem to flourish between the stones. Plumbago and *Solanum crispum* clothe the walls on one side, and the chief treasure of the blue garden, *Echium fastuosum,* provides a forest of great blue spikes all through March. This plant, which is a native of Madeira, and is generally called Pride of Madeira, finds a home among the cliffs on the seashore, but in a cultivated state it is a much more beautiful plant. It is raised from seed, and the plants seem to be at their best about the second year, producing innumerable large feathery spikes of bloom of a very bright blue. There seem to be different strains of it, as occasionally it is merely a dingy grey, and I have never seen it so good a colour in its wild state, nor with such large heads of bloom, so it is to be hoped that this garden variety will be perpetuated, though it is possible that it is merely the soil which affects its colour, in the same way that it affects the colour of the hydrangeas. Even the little fountain in the centre of the garden carries out the scheme of colour, as the water reflects the deep blue sky above, and the fountain itself is made with blue and white tiles, and makes one regret the good old days when tiles, with their patterns in soft harmonious colourings, were used architecturally and let into walls in panels. There are still a few to be seen in the grounds of the Santa Clara Convent, and on the tower of the church, showing that in former days Funchal had probably more architectural beauty than it has to-day.

PRIDE OF MADEIRA AND PEACH BLOSSOM

In April and May the garden seems a feast of flowers in whichever direction you turn your eyes, though there are some good stretches of mown grass to relieve the eye and give a sense of repose. The corridors are clad with roses, among which at this moment the large single white *Rosa lævigata*, with its shiny foliage, is one of the most beautiful. It resembles the Macartney rose, and is often mistaken for it. The plants are seldom entirely without bloom all through the winter, but it is early in April that it becomes a sheet of starry blossoms. Being only half-hardy in England, the climate of Madeira suits it admirably; in fact, I remarked that as a rule it is the roses which are tender in England which thrive best in Madeira. Among the best are the old General Lamarque, which grows rampantly and seems to take care of itself. Its great clusters of snow-white blossoms come in masses in December, and again in April and May. Safrano, Souvenir d'un Ami, Georges Nabonnand, Souvenir de la Malmaison, and Adam, are among the old favourites, though some of the newer kinds of that most beautiful class of roses—Hybrid Teas—seem to take kindly to the climate. It is useless to attempt to grow any Hybrid Perpetuals: they may bloom fairly well the first year, but never again. I have seen good blooms on many of the Hybrid Teas, such as Antoine Rivoire, Madame Abel Chatenay, and others, though never attaining to the perfection of English roses. Possibly the pruning may be at fault, and if the trees were better pruned, better flowers would be the result; but their rampant growth makes them, no doubt, difficult to deal with, and it would be a serious undertaking to cut away all the weak wood from the very large bushes, and certainly the ordinary Portuguese gardener makes no attempt to do so. As a rule, he merely clips the trees, shortening back all the growth equally in the month of January. I believe by a careful system of pruning a succession of roses might be obtained all through the winter, and if, as soon as one crop of bloom was over, the tree was carefully and judiciously cut, a fresh crop could be got in from six weeks to two months.

There are several roses which are to be found in most of the gardens to which I could never put a name: one in particular I can recall, with a beautiful clear, bright pink blossom, touched with a deeper red on the back of the petals, which I frequently admired and endeavoured to get correctly named; but no one knew its name, and at last a friend said: "Why worry about its name? We just call it 'The most beautiful rose that grows'"—and it seemed indeed a good name for it.

CHAPTER V
VILLA GARDENS TO THE EAST
OF FUNCHAL (*continued*)

The Quinta do Til is one of the oldest villas in Funchal, and a description of it is to be found in "Rambles in Madeira and Portugal," published anonymously in the early part of 1826, in which the writer says: "The *Til* is a villa in the Italian style, and possesses much more architectural pretensions than any I have seen here; but it has never been finished, and what has, bears evident symptoms of neglect. The name comes from a remarkably fine *til*, one of the indigenous forest trees of the island, which stands in the garden, *ingens arbos faciemque simillima lauro*: it is, I believe, of the laurel tribe. In the court, too, is an enormous old chestnut, the second largest in the island."

QUINTA DO TIL

The effect of the garden never having been finished is due to the fact that the balustrade of the lower terrace still remains carried out in wood instead of stone, or at least cement and plaster, as was no doubt intended originally. Possibly the death of the original owner caused the property to change hands, and fall into the possession of one who had no sympathy with costly garden architecture. The garden has lost much of its Italian characteristics, as, though not mentioned in the above description, the lower garden was formerly planted entirely with orange-trees, and four large cypresses stood like sentinels near the fountain. Disease killed the orange-trees, as, indeed, it has killed almost all the orange-trees in the island, and the cypresses are also gone, so the garden is now entirely a flower-garden. On the upper terrace the trunk still remains of the chestnut-tree mentioned in the above description; it must have been of gigantic proportions, as the trunk measures many yards in girth. It now supports a single Banksia rose-tree, which is wreathed with its little white starry blossoms in early spring. The chestnut-tree has been replaced by a *Magnolia grandiflora*, which has grown into an immense tree, and is now probably one of the largest in the island. In June, when its large leathery white blossoms expand, it fills the air, especially near sundown, with its almost overpowering fragrance.

The upper terrace is laid out with beds, surrounded by box hedges a foot or more in height, which are filled with an infinite variety of well-grown plants. The garden is very sheltered, and never seems to suffer from the strong, rough winds which those in a more exposed and open situation feel so keenly. Here there comes no rude blast from the east to strip the leaves off the great begonia plants, and their brittle foliage and heavy flower-heads remain unbruised and untorn, while many a neighbouring garden has suffered severely at the hands of a winter storm. Each plant is a perfect specimen in itself, and is the result of many years' care and attention. New-comers to the island are apt to think that in this glorious climate plants are very quickly established, that cuttings will make large plants in at most a few weeks, seeds will spring up in a night—in fact, that gardening is so easy that it is small wonder that gardens filled with plants such as we find here are to be found. Personal experience has taught me that as a rule plants are rather slow to establish, cuttings strike slowly and take a long time to make their roots, especially in the winter months, and the same applies to seeds unless they are sown in early autumn. Once established—say the second year—plants, especially creepers, will make astonishingly rapid growth, but patience is required at first, though well rewarded in the end.

It is evident that this garden is tended with loving hands, and all the necessary alterations and pruning are done under the close supervision of its owners. Their collection of begonias is a large one, and they seem to thrive better in this garden than anywhere else in Funchal, and appear to be in perpetual flower. Pelargoniums of the varieties known in England as Show Pelargoniums, and not of late years much cultivated, new favourites having ousted them from the greenhouse, are here grown into large bushes, many of them five and six feet in height. It is only growing freely in this way that one has any idea of the beauty of many plants which we only know cramped in the narrow area of a six-inch pot. In Southern Italy I remember these same varieties of pelargoniums were grown hanging over terrace walls, and possibly were even more beautiful than when receiving artificial support.

It would again be impossible to enumerate all the plants in this little garden, but it brings to my mind's eye a vision of fuchsias, bouvardias, a beautiful deep mauve lantana, the clear yellow *Linum trigynum*, and hosts of sweet-scented plants, such as verbenas, sweet olives, sweet-scented geraniums, diosmas, and many others.

The lower terrace is almost entirely a rose-garden, the Til garden having always been famous for its roses.

If a few plants of a new rose are imported, the stock can be easily and quickly increased, as the budding of roses, or even grafting, seems an easy matter in this country. The buds take quickly, and the stock may be either that of *Rosa Benghalensis*, which has become naturalized in the island, or any rose which has been proved to have a good constitution may be utilized as a parent. As I have remarked elsewhere, the branch which has been budded is as often as not layered in its turn, and in a few weeks will have rooted, and can be detached from the parent plant; there seems no reason that, once a new variety has been proved to have taken kindly to the climate and soil, a good stock should not be procured and a large group of the same kind planted together, whereby a much better effect is always obtained.

A creeper-clad corridor leads to the group of trees which have given their name to the quinta.

Just above, on the Levada da Santa Luzia, is the gate of the Quinta Palmiera, which takes its name from the large palm-tree which rears its head proudly and stands alone in the grounds. The path leading to the house winds up the side of the hill, through grounds which for many years had been out of cultivation, until the property changed hands a short time ago; but as the ground had always been left in more or less its wild and natural state, it suffered less than if it had been a cultivated garden.

It is a beautiful piece of rocky ground, and on one side a group of *Pinus pinea*, stone, or parasol pines, stand towering over a grand cliff which rises abruptly from the river-bed. In November the rocks are covered with the red spikes of the blossoms of the *Aloe arborescens*, and the effect with the great pines and cypresses beyond is one of indescribable beauty. This is the only villa which can boast of the possession of fine cypresses, and here one realizes the ornament they would be to the island if they were more lavishly planted. The ground near the house is admirably suited for broad terracing, and a splendid effect could be obtained by leaving the cypresses standing out against the distant sea. But the rock being so very near the surface, and the absence of soil, combined with the lack of any means of carting, would make terracing a very serious undertaking.

The grounds contain many very fine trees—among others, a very good specimen of the deciduous cypress, *Taxodium distichum*, which is also called the swamp, or Mississippi cypress, as the whole valley of the Mississippi is clothed with these trees. In summer they are of a splendid deep emerald-green, which gradually turns to a bronze-red colour in autumn, and by December the trees are bare.

At the back of the house there is one of the largest coral-trees in Funchal, and a very large til-tree stands immediately in front of the house.

Among other villas with good gardens, the Deanery, which has long been noted for its fine collection of trees, and the Achada, cannot be omitted. The Deanery, standing in a very sheltered situation at the foot of the Santa Luzia ravine, has proved an admirable trial-ground for trees, shrubs, and plants which have been collected by its present owner. From all parts of the world rare and interesting plants have been brought, and some have been raised from seed on the spot. The following description of the place was written in the early part of the year 1826 by a traveller in Madeira and Portugal, and shows that even in its early days the garden was well cared for:

"To-day we have removed to Deanery, our country-house. The house is a very pretty one. It has not long been built, and, in fact, only a portion of the apartments has as yet been used for residence, but there are more than enough for our accommodation. The situation is delightful—scarcely a quarter of an hour's walk from Funchal, and enjoying, from its comparative elevation, a beautiful view down the valley to the city (which, though so near, is scarcely visible from the orange-trees and cypresses that embower us), and to the bay and coast and the blue Desertas beyond. Close on the west is the Santa Luzia ravine, the farther side of which rises to a considerable height, its cliffs terraced, in the way I previously described, into little gardens and vine-grounds, and crowned by the trees and trellises of the Achada Quinta.

"Our great luxury, however, is the garden. It is one of the largest and most beautiful in the island. A spacious vine corridor runs round nearly its whole extent, under the green arches of which in summer, you may either ride or walk in coolness, while the interior space forms a 'leafy labyrinth,' in which trees and shrubs, flowers and fruits of every clime are here crowded into a wilderness of shade and beauty. The higher part of the ground, upon which stands the house, is elevated considerably above the rest, and is divided from it by a terrace of considerable height. This circumstance is of very happy effect for the beauty of the garden: it in a manner doubles its extent, and multiplies its variety; while the wall of the terrace, in some parts nearly twenty feet high, affords an admirable field for every species of tropical creeper, to luxuriate, as it were, at full length, and to put forth its leaves and blossoms to the sun, in all the fearlessness which such a climate and aspect justify.

"Above the house the ground rises another step, and the boundary of the garden here is a wall of native rock, which is already half veiled with the trees and trailing plants interposed to relieve its ruggedness. The freshness of the scene is completed by the tanks, always copiously supplied with running water, and which a little trouble might, I think, bring into play as fountains."

Across the ravine, but at a very much higher altitude, stands the Achada, in a commanding position on, as its name implies, a stretch of level ground. The road leading to it from the town, known as the Caminho da Sao Roque, as it eventually leads to the village of that name, is almost as steep as the Mount Road, and a very pretty view of the town is visible between its creeper-clad walls, with the picturesque church and tower of Santa Clara in the distance. The Achada has also long been famous for its garden and grounds. It formerly belonged to an English family, who probably planted most of the rare trees, palms, and Dracænas, and the large magnolia-trees for which it has become famous. The property then changed hands, and for some years belonged to a Portuguese family, but is now again in English hands. The following is by the same unknown author of the above description of the Deanery in 1826: "The English merchants all have mansions in the city, but they commonly live with their families in the country-houses in the neighbourhood of it. To-day we have been returning visits, which has taken us to some of the finest of these quintas. One of them is the Achada. The situation is delightful: it stands on a level, the only one in the environs, just above the city, and thus enjoys an advantage in respect to surface possessed by no other. The grounds are extensive, rich in fruits and in flowers, and surrounded by alleys of vine trellises. These vine corridors, as they are called, are common to all the gardens, and in summer, when the plant is in leaf, must be peculiarly grateful."

ON THE TORRINHAS ROAD

CHAPTER VI
THE PALHEIRO

About an hour's ride from the town, at a height of some 1,800 or 2,000 feet, is the Palheiro, formerly known as Palheiro de Ferreiro (Blacksmith's Hut), the principal country place in the neighbourhood of Funchal, belonging to the same owner as the Quinta Santa Luzia. The road leads past many smaller villas, whose gardens have most of them fallen into decay, and only undergo a hurried process of tidying when their Portuguese owner comes to spend a few weeks away from the summer heat of Funchal.

Palheiro was not entirely laid out by its present owner, though the grounds have been very much enlarged and improved, and the house itself, having been destroyed by fire a few years ago, has been lately rebuilt. Some letters from Madeira, written by J. Driver and published in 1834, give the following interesting account of Palheiro, which in those days belonged to the family of Carvalhal.

"The grounds of Senhor Jose de Carvalhal are the finest in the island, possessing a level surface, which is very difficult to be met with here to any extent. This place was recommended to us for our first ride into the country, and after some delay in making choice of the ponies and *burroquieros* that we intended afterwards to patronize, we made our way eastward out of the city. Crossing a bridge over the deep bed of a river, we saw the ruins created by the great flood in 1803, when several hundred inhabitants were swept into the sea. We now ascended a steep and narrow road for a distance of two or three miles, passing several of the merchants' houses, from all of which there is a commanding and beautiful view of the city and the bay. The Palheiro, lately the residence of Senhor Carvalhal, by far the richest *hidalgo* of the island, has been confiscated by the Miguelite Government. Senhor Carvalhal himself had some difficulty in effecting his escape; however, he got on board an English vessel in the bay, and is now residing in London. Upwards of 700 pipes of very choice and old wine were at once taken from his cellars, and sent to Lisbon to be sold on Government account. The house was ransacked, and his grounds are now (though this is of recent occurrence) fast going to ruin. There are a few soldiers stationed near the house to prevent any material damage, and these are now the only persons to be seen on this once splendid estate. The park, if we may so term it, is more in the English style than we expected to find it; but when we came to the orange, lemon, pomegranate and shaddock groves, which are in fine foliage and planted in the best order, we at once saw the effect of these Southern climes. The flower-gardens, though not abounding in that variety we might expect, are well arranged, but begin to show more of the 'fallen state' of things than the other parts of the grounds. The house itself is not on a large scale, yet it is built in good style and keeping with the place, as well as the chapel, which is a neat edifice at a short distance from the house. Senhor Carvalhal used to employ more than two hundred men on the estate, for the purpose of keeping it in order. He was a kind landlord, and much respected throughout the whole of the island. Let us, then, hope that Portugal will soon have a fixed Government, and that Senhor Carvalhal will return to his country, and again have the pleasure of enjoying his estates."

The hope here expressed was fulfilled, and the family continued to live there until the estate changed hands and became the property of the present owner, in 1884.

Although on first acquaintance it is true that the grounds suggest those of an English park, possibly from the welcome presence of turf, and also from the fact that at that high elevation the deciduous trees are leafless throughout the winter, like Mr. Driver, we shall very soon discover many trees and shrubs that could not be grown in even the most southern parts of England, though many English shrubs and flowers flourish in the warmer climate.

There are two roads leading from the outer gate to the house. The lower road winds through a long avenue of camellia-trees, whose branches in January and February are laden with their single, double and semi-double blossoms, ranging in colour from pure white, through every shade of pink, to deep red. Along the higher road, beneath the trees, broad stretches of the deep green leaves of the *Amaryllis belladonna* give promise of beauties to come. In summer all trace of their foliage vanishes, and early in September the deep red stems and sheath of their flowers begin to appear. By the end of September their blush-coloured flowers will have developed; and so profusely do they flower that all through October in these higher regions the land is transformed by their rosy loveliness. Like the garden of Santa Luzia, Palheiro has been made the trial-ground of many an imported treasure, and many which did not flourish in the warmer and drier regions have succeeded admirably in the cooler and damper air of the hills.

The flower-gardens certainly show no signs of the "fallen state of things" under their present ownership, and a small enclosed garden a short distance from the house is a perfect treasure-house; though naturally at its best in spring and summer, it is never devoid of flowers. Here English daffodils, pansies, and polyanthuses grow side by side with many a bulb and plant which will just *not* stand the rigours of our English winters. The large-flowered violets, Princess of Wales and other varieties, flower in their thousands from November till April, with blooms so large that they suggest violas more than violets. Freezias and ixias have seeded themselves in the grass slopes of this little favoured garden, where the beds are enclosed by trim box hedges. At the corners or angles of the beds the box is cut into all sorts of fancy shapes, such as pyramids and ninepins. In the beds grow large masses of the pale yellow sparaxis, anemones of every shade, single, semi-double, or double, and the graceful little *Cineraria stellata*, in an infinite range of soft colouring. Or a whole bed is devoted to the deep purple *Statice*, the beautiful white *Alstrœmeria peregrina*, or some other chosen flower which gives a definite note to the colour scheme. In March two fine specimens of *Magnolia conspicua* are covered with their cup-like white and lilac blossoms, and stand out in sharp contrast to the deep emerald-green of the *Araucaria braziliensis*, which forms an admirable background to them, and is in itself one of the most beautiful of all trees. Near the magnolias a large shrub of *Cantua buxifolia*, with its bright red tube-like blossoms hanging in graceful bunches, provides a brilliant patch of colour. The lilac *Iris fimbriata*, with its branches of delicately veined flowers, seems to flourish in the shade, and though its individual blossoms are short-lived, they are so freely produced that for many weeks in the late winter and early spring the plants remain in beauty. One could linger for many a long hour in this peaceful spot, resting in an arbour completely formed of the clinging, twining *Muhlenbeckia*, which has grown into so dense a thicket that it provides welcome shade and shelter, or wandering from one little terrace to another, examining the endless treasures the beds contain; for, as the garden has a wealth of flowers all the summer, there are many things which, from being out of flower, might pass unnoticed.

Great beds of *Azalea indica*, and trees of different varieties of mimosa, bending under the weight of their golden blossoms, remind one that this is no English garden, while glades and banks show long vistas of white arum lilies, as *Richardia* or *Calla Æthiopia* are commonly called. Here these African lilies, which are also called lilies of the Nile, are completely naturalized, and bloom continuously for at least five or six months of the year.

A deep dell, shaded by mahogany and other trees, has provided a home for the tree-ferns of Australia, New Zealand, and Africa, and in some twelve or fourteen years they have made such astonishingly rapid growth that the little ravine is suggestive of the celebrated fern-tree gullies of Australia or Tasmania. The ivy, which hangs from tree to tree in long ropes, replaces the lianes of a tropical forest, and the banks are clothed with woodwardias and other ferns, while a few of the rarer native wild-flowers, such as the monster buttercup, *Ranunculus grandifolia*, and the giant fennel, have been introduced, and are thoroughly in keeping with their wild and natural surroundings. A path winds down the little valley following the bed of the stream, and on emerging from the deep shade of the fern-trees, broad masses of naturalized plants are revealed with every turn of the path. On a grassy slope, over which tower two or three grand old stone-pines, thousands upon thousands of golden lupins have sown themselves. A single specimen of a plant may often hardly be regarded or considered worthy of notice, but the same plant, when seen in great masses, may call forth universal admiration because of the wealth of colour it provides. In summer the agapanthus will send up innumerable heads of clear blue flowers, while the little *Fuchsia coccinea* seems to flower bravely at all seasons of the year. In order to show that even in this favoured land it is possible to have failures in the gardens, and importations from other climes do not always succeed, some rhododendrons, even the common *ponticum*, were pointed out to me as never having made themselves at home, and in a shady corner hundreds of our English primroses had been planted, but had pined away and died.

In another part of the garden the beautiful rhododendrons from Java are being given a trial; but possibly, just as the climate is too hot for the hardier varieties, it may prove too cold for those from tropical regions. The variety known as *arboreum*, with its large heads of deep crimson flowers, appreciates the climate, and has no spring frost to cut its blossoms, which so often mars the beauty of this very early-flowering rhododendron in England, where, for this reason, it only succeeds in sheltered situations. The large white variety, which is commonly called the Himalayan rhododendron, though, more correctly speaking, it is known as *Edgeworthii*, flourishes here. It was introduced from Sikkim to Europe in 1851. It is a shrub of somewhat straggling growth, with large wide-open pure white flowers, sometimes tinged with yellow or blush; they are produced in small clusters, not more than three or four together, and diffuse an overpoweringly strong scent.

Among new importations are a collection of Japanese cherry-trees, including the beautiful and graceful weeping variety and some of the double-flowered kinds, also the deep pink plums, which should all prove a success, as in the little flower-garden described above a large double-flowered pink peach-tree is the pride of the garden when in blossom.

Besides these so-called fruit-trees, which are only cultivated for their beautiful blossom, and bear no fruit, many fruit-bearing cherries, plums, and peaches have been planted in the more prosaic part of the garden; but the stone fruit, is only a partial success. The peaches seem to deteriorate when the trees have been more than a few years in the island. Possibly the pruning is at fault, or the fruit forms and ripens too quickly; and when the plum-trees are laden with fruit, a *leste*—the cruel, hot, scorching wind which the natives dread in summer—will blow for a few days, and shrivel the fruit and spoil the whole crop.

The orange-groves have vanished, destroyed by disease, which gradually spread from Funchal throughout the island, up to the higher land. The lack of enterprise common to all Southern races being a marked feature among the Portuguese, no combined effort was ever made to check its devastating progress.

The garden has no definite boundary, no unsightly garden fence, which is the stumbling-block of so many gardens. One can wander down through the pine woods, or up the hill, where, looking west, the whole bay and town of Funchal lies spread out like a map before you, or, looking east, the distant islands seem to provide a never-ending variety to the view. Sometimes the islands look dark against the sky, which means storms ahead; or sometimes they are wrapt in a soft haze, which means a promise of fine weather; or the setting sun may have caught and kissed them with her last departing rays, and made them blush a rosy-pink, and one is tempted to linger and watch the light gradually fade; but it is time to turn homewards, as in these Southern latitudes twilight is all too short, and darkness descends quickly over the land.

CHAPTER VII
CAMACHA AND THE MOUNT

The road past Palheiro leads, through pine woods and long stretches of yellow broom and golden gorse, to the little mountain village of Camacha. Probably the village has become noted for its flowers from the fact that many English people, in the days when travelling was not so easy, used to make this place their summer-quarters, instead of returning to England, as they mostly do in these days of quick travelling.

WISTARIA, QUINTA DA LEVADA

One garden I can recall which, though now neglected, still shows how it was once well cared for. Though the turf is no longer mown, and the box hedges have lost some of their trimness, the beds are still full of what were once treasured plants. The rose-garden no longer sees the knife of the pruner, but the trees grow and flower at their own sweet will, in careless disorder. It is a very lovely disorder, but it is always sad to see a garden once tended with the greatest care fall into other hands, who know nothing of the art of gardening. In spring the garden was full of jonquils and narcissi, and later on sparaxis and ixias. Near the house great bushes of *Romneya coulteri* were covered with their delicate white poppy-like flowers in summer. The plant seemed to have become thoroughly established, and threw up suckers in all directions, even through the paths of hard-beaten earth. From the grounds there are lovely views of the sea; and probably the garden looks its best when the agapanthus sends up its flowers in hundreds, and the hydrangea bushes are laden with their bright blue blossoms—as blue as the sky above or the sea below; or, again, in October, when the belladonna lilies are flowering in their thousands.

I think the love of gardening must have spread from these English gardens to the native cottage gardens. The English probably encouraged the cottagers to cultivate their plants, as from these little gardens come all the flowers which are to be bought in Funchal. A few flower-sellers will trudge seven long weary miles down to the town, nearly every day of the week, with a heavy basket of flowers on their heads, which they have collected from many a cottage garden. Naturally these flowers are not of the best, and it is very much to be regretted that some enterprising person does not start a shop or garden where cut flowers and plants could be bought. Many a time have I been asked where, in this land of flowers, good cut flowers can be procured, and the answer has had to be "Nowhere." Would-be purchasers have to satisfy themselves with the contents of these baskets which are brought to the hotel and villa doors, and their contents are far from satisfactory. Beyond arum lilies, violets, and irises, a few indifferent daffodils and poor roses, there is little to be got. The women will complain that they have not a large sale for flowers, and it is in vain that I have told them that the real reason of it is that their flowers are so poor. Nosegays of a mixture of a dozen flowers, in as many colours, naturally find no market; but good flowers, I feel sure, would have a large and ready sale at reasonable prices.

The little gardens at Camacha are gay with common flowers: large bushes of white marguerites and trees of the early-flowering red *Rhododendron arboreum* give colour to the village even in early spring, and in summer it is naturally much more flowery. On every bank and hedgerow grow bushes of hydrangeas, with their flaunting blue blossoms, while great clumps of belladonna lilies transform the whole landscape, and the country seems to blush a beautiful rosy-pink.

The road between the two most popular summer resorts, Camacha and the Mount, runs through pine woods and long stretches of golden gorse to the Pico d'Infante, from where a very fine panorama of the Bay of Funchal is to be seen by turning aside a few yards from the road. Just beyond this point the path strikes the Caminho do Meio, another steep road leading down to the town. Near the eucalyptus and pine groves is the Quinta Bom Successo, one of the most beautiful of the outlying properties, which, from its elevation, escapes the summer heat, while its sheltered and sunny aspect makes it a pleasant residence through the winter months. The large grounds extend to the edge of the ravine, and a view of surpassing loveliness is suddenly brought before one at the very end of the terrace. The river roars and tumbles below, and the ragged cliffs throw deep mysterious shadows, while the more distant hills are wreathed with light transparent mists. The sides of the cliff have been transformed into a wild garden, as many plants have strayed from the garden proper, and have either seeded themselves or been cast over the precipice as discarded plants, where they have taken root and clung to life in some cranny between the stones. Within the grounds a rocky bank is covered with great stretches of the red *Aloe arborescens*, blue agapanthus and vast clumps of belladonnas, all growing in careless profusion. The garden has long been noted for its orchid-houses, where plants have been brought from all parts of the world, and also for the pine-houses, from which hundreds of pines are cut annually. Showing that, though at a comparatively high altitude, the garden is sheltered and warm, two natives of Burmah, the giant honeysuckle, which in May is wreathed with its strong-scented trumpets and the Burmese rose, both flourish, and in a few years have made astonishingly rapid growth.

The road to the Little Curral leads past a grove of *Mimosa cornuta*—which is smothered with its fluffy balls of yellow blossoms in early spring—to the valley itself. Every fresh turn of the steep zigzag path opens out fresh views, and at every step a new fern or little wild-flower is to be seen nestling between the damp mossy stones. Down near the bed of the river, which tumbles over great boulders in a roaring torrent after heavy autumn or winter rains, a large colony of arum lilies begin to unfold their pure white flowers in November, and continue in one unceasing succession until the late spring or early summer. The path winds up the opposite hillside, through a group of peasants' huts, where yapping dogs and begging children for a few minutes mar the harmony and repose of the scene, and then again the path enters another silent valley, until the little village of the Mount is reached. A colony of countless little quintas, which have sprung up under the shelter and protection of the Church of Nossa Senhora do Monte, has of late years become a more favourite summer resort than Camacha. The air may not be quite so pure and cool, but the proximity of the town and the convenience of the funicular railway are, no doubt, responsible for its growing popularity.

The principal villa, the Quinta do Monte, formerly owned by an Englishman, has large grounds, planted with many rare trees and shrubs. The property has changed hands; the house is no longer inhabited, and the garden is falling into decay. As the grounds were always more pleasure-grounds than actual flower-gardens, it has suffered less than a smaller garden, which misses the personal care of its owner. The camellia-trees are an immense size, and have out-grown the little garden centring in a sundial, in which they were, no doubt, originally planted as small shrubs in beds with neat box hedges. Here are to be found tree-ferns, long rows of agapanthus, and a great plantation of mimosa-trees, which is quite a feature in the landscape in early spring, when they are laden with their balls of yellow blossoms.

In every direction in this district large clumps of the foliage of the belladonna lilies are to be seen in winter, on every bank, in every little garden: giving promise of their glories to come in the waning summer months. But in the grounds of Quinta da Cova they are probably to be seen at their very best, as here they have been more collected together, and broad stretches of them carpet the ground in thousands, beneath the chestnut-trees. I remember once hearing a traveller remark, who had passed through Madeira in August, on his way to the Cape, and returned again early in October, that when he first saw the island "it was all blue," alluding to the effect of the agapanthus and hydrangea blooms, and when he returned it had changed, and was "all pink," from the masses of belladonna lilies.

CHAPTER VIII
A RAMBLE IN THE HIGHER ALTITUDES

The Church of Nossa Senhora do Monte is the starting-point of many an expedition made by those who have a wish to see more of the beauties of the island than can be done within the restricted area of Funchal. Should the Metade Valley be the point chosen, or the bleak Pico Ariero, with its enchanting views, or should the traveller be bent on a longer tour, and be proposing to make the little village of Santa Anna his headquarters for seeing the beautiful scenery of the north side of the island, the road up to a height of some 4,500 feet will be the same. Gradually the steep path winds its way through the fir woods, which in the early morning while the dew is still on them, exude a delicious aromatic scent, and the bushes of the little red *Fuchsia coccinea* and *Rosa Benghalensis*, with its small double pink flowers, and the clumps of belladonnas on the banks, which at first give the landscape the appearance of a ruined garden, are left behind, and the vegetation changes completely.

The pine woods consist chiefly of plantations of *Pinus maritima*, or *pinaster*, which have been planted for practical purposes, and have replaced the more beautiful chestnut woods, which were wantonly destroyed. These pines, being of rapid growth, are soon cut down, and provide timber for firewood, garden and vine trellises—in fact, are strictly utilitarian. The roots and stumps are burnt on the ground, and then possibly a crop of some sort is sown before the fresh pine seed is put in. This system has been the means of saving some of the more valuable and beautiful native trees, which at one time were ruthlessly felled; and even the forests in the interior, so necessary for the preservation of the water-sources, were threatened with destruction. Interspersed with the plantations of pine-trees are broad stretches of the common broom, which is sown extensively on the mountain-sides, either for the purpose of being cut down for firing, or to be burnt on the spot every five or seven years to fertilize the ground, and cause it to produce a single crop of wheat or batatas. The twigs and more slender branches are commonly used for making into faggots, and numbers of country-people, especially young girls and children, within reach of Funchal gain a scanty and hard-earned living by bringing daily into the town, often from great distances, bundles of *giesta*, as the natives call it, to be used for heating ovens and igniting the larger firewood. Doubtless the species was originally introduced into Madeira, though it is proved to have existed there for over 150 years, and now is so extensively diffused that it appears to be perfectly naturalized; in spring it floods the mountain-sides for miles with seas of its golden blossoms. The very fine and delicate basket-work peculiar to Madeira is manufactured from the slender peeled twigs of the broom.

Gradually ascending to the higher altitude, those who can tear their eyes away from the beautiful view of the Bay of Funchal and the curiously shaped hills above the villages of Santo Antonio and Santo Amaro will notice that by the roadside, in the moisture exuding from between the rocks, the innumerable ferns and the common foxglove, which at a lower altitude were so abundant, will gradually vanish. The myrtles, formerly so fine, are now unfortunately becoming almost scarce, owing to their injudicious destruction for ornamenting churches and adorning religious processions, after a height of 3,000 feet are no longer to be seen, and the country gradually becomes barren of vegetation. Rocks of basalt and red tufa appear, and the long sweeps of turf are only broken by large bushes of a heath, called, I believe, *Erica scoparia*, which, from being constantly eaten off by the mountain sheep and goats, gets a curiously distorted and stunted growth, though they eventually attain to a large size, and have such venerable-looking stems that they are suggestive of the dwarfed trees of the Japanese. Then comes the region of the *Vaccinium Maderense*, or *padifolium*, which varies in appearance according to the season. In winter it has crimson foliage, then it bears waxy bell-shaped blossoms, and in autumn is covered with almost black berries. From the situation in which it grows, exposed to the full blast of the north wind which sweeps over that stretch of country, it also has a bent and distorted appearance; and the dampness of the air—as, more often than not, at this altitude a white mist envelops the land—causes its stems to be covered with the *Usnea* lichen, which waves from one tree to another like masses of long green hair.

A turn in the road, at an altitude of some 4,800 feet, just beyond the rest-house at the bleak spot known as the Poizo, reveals a grand chain of mountains, with deep ravines running down to the sea. The traveller's path will wind, in zigzag fashion, down the steep mountain-side, and gradually the *Vaccinium* will be left behind and the beautiful ravine of Ribeiro Frio is entered—thickly wooded with many varieties of the laurel tribe, which in their turn have their stems clothed with lichen.

To collectors of wild-flowers and ferns these mountain expeditions are a never-ending joy, as, according to the different seasons of the year, innumerable treasures are to be found. A ramble along the many *levadas*, or water-courses, will well repay the collector, as at all seasons, ferns, mosses, lichens, lycopodiums, and hosts of other moisture-loving plants, are to be found; while in June and July, when the wild-flowers are in all their glory, many rare and interesting plants will appear. The levada which runs through the Metade Valley was formerly the home of the *Orchis foliosa*, the orchis known everywhere as peculiar to Madeira, and its bright purple spikes brightened the dense masses of green. Of late years the plant has become scarce, probably ruthlessly uprooted by passers-by, or in order to be offered for sale in the town of Funchal. In describing this beautiful ravine, over which towers Pico Ruivo and the Torres, both some 6,000 feet in height, Miss Taylor, who was a great authority on native ferns, says: "Many rare and beautiful ferns will be found, growing both close to the running water and on the mountain-sides above the levada. *Trichomanes radicans* and *Hymenophyllum Tunbridgense* grow in great abundance; also *Acrostichum squamosum, Pteris arguta, Asplenium umbrosum, Woodwardia radicans,* and numberless others. Lichens of every sort and mosses—*Lycopodium suberectum* and *Selaginella Kraussiana*—seem to fill up every available space and crevice, and engage the hands and delight the mind of the collector."

The more arid path to Ariero will not provide such treasures for the collector, who must content himself with the views of surpassing loveliness down to the deep, wooded ravines, which as the shadows begin to lengthen after midday, grow more mysterious-looking, getting grander and more beautiful as their deep blue turns to purple; and gradually the haze, which is certain to come before nightfall, fills the valleys and blots out the sea beyond. The rare orchis *Goodyera macrophylla* is said to be found in this district, with its beautiful pure white spikes, and here and there thickets of a low-growing indigenous, mountain ash, which in September bears fragrant white flowers, to be followed by brilliant scarlet berries in early winter.

From just beyond the rest-house at the Poizo a long turf ride of some four or five miles leads to the Lamaceiros, and is a welcome relief after clattering over the eternal cobble-stones. A long round, over country where seas of golden gorse, when it is in bloom, delight the eye and nose and make a beautiful foreground to the enchanting views, leads eventually past wooded glens, either over the Portella down to the village of Santa Cruz, or through the village of Camacha back to Funchal. A levada near the reservoir at the Pico d'Assoma is again rich in ferns, and Miss Taylor says: "The lover of ferns will perfectly revel in the wealth of lovely *Hymenophyllums* which clothe the stems of old laurels: here and there a mass of rock, perfectly cushioned with *Hymenophyllum Tunbridgense;* here and there a carpet of *Hymenophyllum Wilsoni* and *Davallia Canariensis* and *Polypodium vulgare* growing in masses on the trees. *Nephrodium Oreopteris* here grows in great abundance, the one place besides Pico Canario where it is to be found in Madeira. *Nephrodium Fraenesecii* and *Nephrodium dilatatum* here grow very large and perfect. The levada is fringed with *Asplenium monanthemum*, *Cystopteris fragilis*, and countless treasures. In July the *Orchis foliosa* blooms in great spikes of bright mauve. In this neighbourhood *Acrostichum squamosum* and *Trichomanes radicans* grow well."

Probably nearly every levada in the island would repay exploring, but some are very inaccessible and require a steady head. One of the most beautiful is certainly that of the Fajao dos Vinhaticos, which could disappoint no one, and can be seen by staying at the village of Santa Anna, or, better still, at the engineer's house on the levada itself.

On the north side of the island the vegetation is mostly the same. The rough and precipitous path which winds through the Boa Ventura Valley up to the Torrinhas Pass is clothed mostly with trees belonging to the laurel tribe. From the Pass itself some of the grandest views in the island are to be seen. The grandeur of the rocks and the splendid vegetation, the profusion of ferns and wild-flowers, hare's-foot ferns hanging in long fringes from the stems of the evergreen trees, the variety of lichens, some of a deep orange colour, make the long ascent an endless source of delight to lovers of Nature, and, provided the weather is fine and the valleys free of mist, I know no more beautiful expedition.

If the traveller is returning to Funchal, he will gradually descend from this high altitude (close on 6,000 feet), down past the Church of Nossa Senhora do Livramento (Our Lady of Deliverance), through the valley of the Grand Curral, up the steep zigzag road opposite, and back to Funchal through the village of Santo Antonio. The region of the laurels and ferns, dripping with moisture, is left behind when the traveller turns his back at the top of the pass on the beautiful Boa Ventura Valley, and he will gradually return to the region of the heaths, pine woods, broom, and gorse.

When the village of Santo Antonio is reached, a marked change in the vegetation will be noticed. There are many Spanish chestnut-trees, whose fruit, being very popular with the natives, is sold in bushels in the town in autumn and early winter; and, the district being a very warm one, on the banks and in the hedgerows by the wayside the prickly-pear, agaves, and cactus will begin to appear, while large clumps of pelargoniums, sweet-scented geraniums, and lantanas have strayed from gardens and sown themselves in every direction. In April the beautiful *Ornithogalum Arabicum*, bearing its white starry blossoms with jet-black centres, may be seen growing wild, and I have been told that the pure white *Lilium candidum* is to be found in a wild state, though I have never come across it myself.

Between Santo Antonio and Santo Amaro the earliest strawberries which are brought into the market in Funchal are grown, making their appearance in favourable seasons late in February, though at that season they have little flavour, and generally only find favour in the eyes of the tourists, who are attracted by their inviting appearance as they are offered for sale in little fancy baskets. If some enterprising person would make some experiments with growing the plants on rather steep banks or slopes, as I have seen done elsewhere in temperate climates, in order that the plants may get the full benefit of the sun, I feel almost certain that far better early strawberries could be obtained: the sun would draw out that watery flavour from which they suffer. But it is always hard to induce a cultivator of any nationality to try new methods, and in vain one preaches, and is only met with pitying looks of incredulity and the remark that the crop, whatever it happen to be, has always been grown in the same way, however bad a way it may be, by the present owner, his father and his grandfather before him, and what was good enough for them is good enough for him.

There are more vines grown here than in any other neighbourhood, though, in consequence of the numerous attacks of disease—two scourges having several times threatened to completely destroy the vineyards: the dreaded Phylloxera insect, which attacks the roots of the vines, and also *Oïdium Tuckeri*, which settles on the leaves and fruit—together with the depression in the wine trade, vines are far less grown than formerly. Being trained over corridors—or *latadas*, as they are called in Madeira, *pergolas*, as they would be called in Italy—the effect is not only very pretty, but seems practical, as, being at a sufficient height from the ground, a labourer can work underneath them, and it is not uncommon to see another crop growing between the vines, though this practice of overstocking the ground is no doubt responsible for the failure of many a crop. The vines are pruned in February, though not to any great extent, and in April start into growth, and soon clothe the corridors with fresh young leaves and long twining tendrils. The flowers come in May, and by August the vines are laden with fruit ready for the harvest, which in early seasons begins in the lower regions late in August and continues, according to the altitude, until October.

The cultivation of vines and bananas, which were also grown at one time to some considerable extent, has been almost entirely replaced by that of sugar-cane, which, in consequence of the current rate fixed by the Government being a very high one, is at the present time a very profitable crop.

The cultivation of sugar-cane in the island dates from very early times, as in Cadamosto's Voyages he writes that he visited the island in 1445, only twenty-six years after its discovery, and says: "Zargo caused much sugar-cane to be planted in the island, which has done well, and from which they have made sugar." Mr. Yate Johnson says: "The cane is thought to have been introduced from Sicily about 1425, at the instance of Prince Henry. The first plantation was made on the site of the Cathedral, and did so well that the cane spread to other localities. Matters proceeded so rapidly in those days that in 1453 a mill was erected for crushing the canes by means of water-power.... Prince Henry was a good business man, and knew what he was about in making a bargain, for it was stipulated that he should receive one-third of all the sugar produced. Another stipulation was that the mill was to be placed where it would not be an annoyance to others, a regulation which, it is to be regretted, is not enforced at the present day. It is not known where this first mill was built, but it is more likely to have been in Funchal than anywhere else." By 1498 the production of sugar is said to have increased to a very large extent, and then came troubles in the trade. The introduction of the cane to the West Indies and its extensive cultivation there caused increasing competition in European markets, and led to a heavy fall in price; but notwithstanding this, the cane continued to increase in Madeira, and by the end of the fifteenth century a large number of slaves were employed, both as labourers on the land and in the mills, which by now had increased in number to 120, on the southern side of the island.

Early in the sixteenth century disease came, in the form of a grub which eats into the cane, and the plantations suffered severely from its ravages, though many attempts were made to check its depredations. Possibly this, combined with the abundant production in the West Indies, caused the sugar-growing in Madeira to become so unprofitable that the mills dwindled down to only three in number, and the cultivation of vines for a time reigned supreme. This, in its turn, received so severe a check through the grape diseases in 1852, that the cane was once more restored to favour and again extensively planted. The cultivation increased, and new crushing machinery was imported from England; steam-power replaced the more primitive methods of water-power, or working the mills with bullocks only. After the revival, for a time the cane was only used for its juice, to be distilled into spirit (*aquardente*), but gradually, new sugar-making machinery having been imported, its manufacture was resumed and continued, until it has now reached the vast amount of about 2,500 tons per annum.

Different kinds of cane have been introduced, and if the cultivation is to be continued at the present enormous extent, artificial manures will have to be largely employed to prevent the soil becoming exhausted. The cane—I may say luckily—cannot be grown above an altitude of about 1,700 feet, or it would seem as if there would be no end to its cultivation, which by no means adds to the beauty of the island, and to my mind is an unsightly crop.

RED ALOES

CHAPTER IX
A RAMBLE ALONG THE COAST

The vegetation along the seashore is naturally very different to that at a higher altitude. Wherever it has been found possible, the ground has been brought into cultivation, even up to a height of 2,500 feet. Pressed by the ever-increasing population, and the consequent need of more food for more mouths, the country-people are continually bringing into cultivation fresh patches of ground. No minute piece seems to be wasted, and many an odd corner and neglected patch which, from its steepness or the poor quality of soil, escaped cultivation in years gone by, being rejected as incapable of bringing any return for the vast labour which has to be applied to it in the first instance, has been, as it were, pressed into service of late years. The larger expanses of cultivated ground have been utilized for the profitable and ever-increasing sugar crop, and these tiny terraces, when the stones have been dug out or the rock blasted, and walls built to support possibly only a few square yards of the levelled ground, will grow a scanty crop of some article of food. Thus bit by bit the cultivation has crept up the hills, and has done much to mar the beauty of the island. The peasants are very primitive in their modes of cultivation, and as long as the ground receives occasional irrigation during the hot, dry months, and the surface is roughly broken with their native hoe, it is all they consider necessary, and are strongly averse to every kind of innovation. It is small wonder that even in such a climate the crop suffers; the earth becomes impoverished and the vegetables produced are of a most inferior quality. Their principal root crops are the ordinary potato; the sweet potato (*Batata edulis*), a plant of the convolvulus family; and the *inhame*, a kind of yam. The sweet potato is one of their staple articles of food, and the native appears to consume an inordinately large quantity of *batatas*. The tuberous roots yield three or even four crops annually. In situations where the ground can be kept constantly so supplied with moisture as to be in a swampy condition, the *inhame* (*Colocaria antiquorum*) is grown even up to a very high elevation, some 2,500 feet. It is quite different to the West Indian yam, and belongs to the arum family; indeed, its leaves at once suggest those of arum lilies, only the roots are edible. These are another most important article of food. Other crops are haricot beans, the ripe seeds of our French beans, whose young pods are nearly always in season; but with the Portuguese it is the ripe seeds (*feijoens*) which are most valued for making their *sopas*, or vegetable soups. Lupines, lentils, and the chickpea (the *grao de bico* of the Portuguese), broadbeans, and peas, come into market in the winter months, but are of very poor quality and singularly tasteless, even when gathered young, which it is very difficult to persuade the peasant cultivator to do. That they need not be poor in quality and flavour, if more pains were taken in their cultivation, is proved by the fact that in private gardens where fresh seed is imported from England or America excellent peas can be grown. Another most important article of food is derived from several varieties of the pumpkin tribe, and in summer over every trellis, and even on the straw roofs of the peasants' huts, the gourd-bearing plants are trained, and their *aboboras*, as they are called, are carefully tended. Mr. Lowe writes: "For at least six months in the year (August to January) the *aboboras* constitute almost one-third of the daily nourishment of all classes; and from their facility of combination by boiling with fatty substances, together with their large supply of saccharine, besides their farinaceous material, afford a most nutritious food, evinced by the surprising muscular power of the Madeiran peasantry." The pearshaped, green, wrinkled fruit called pepinella (*Sechium edule*), or chou-chou by the English, is not unlike a cucumber, and yields a constant supply in the winter months. Spinach, cabbages, and cauliflowers are, I believe, only grown for the requirements of the English, and to provision the passing ships, and with these the list of vegetables closes—and somehow is a disappointing one—and many an English person longs for the fresh vegetables from a home-garden.

Nor is the list of fruits a long one. The orange-tree has practically died out. The apathy of the native made him consider the task of fighting the disease called scale, induced by an insect, too arduous a one, as constant washing of the trees is necessary to prevent its ravages; and he remained content to see all the orange-groves disappear, and the fruit is now imported from the Azores, Portugal, and even South America. At one time, we are told, the vast banana plantations gave quite a tropical aspect to the gardens about Funchal; they have been largely replaced of late years by sugar-cane, and are no longer so extensively cultivated as the facilities due to cold storage on ships flooded the European market with bananas of the West Indies. Several varieties are grown, but the fruit of the silver banana, a tall growing kind, is most prized and fetches a higher price than that of the dwarf *Musa Cavendishii*. In an old account of Madeira, printed in Astley's "General Collection of Voyages and Travels," the following curious account of the plant appears: "The banana is in singular esteem and even veneration, being reckoned for its deliciousness the forbidden fruit. To confirm this surmise they allege the size of its leaves. It is considered almost a crime to cut this fruit with a knife, because after dissection it gives a faint similitude of a crucifix; and this they say is to wound Christ's sacred image."

ALMOND BLOSSOM

Sufficient lemons and citrons are grown to supply the requirements of the island. The custard apple, *Anona cherimolia*, ranks high among the island fruits, and is hailed with delight when it first appears in the market in late autumn. In common with the guava, it was originally imported from America; while the mango, whose fruit leaves room for much improvement, came from India. Guavas are extensively used, either uncooked, stewed, or possibly in the most favourite form, made into a clear, transparent jelly. The loquat bears abundantly, and as it is very readily increased from seed, has become a very common tree, though I do not consider the fruit to be as good as those of the Italian loquats. The pittanga, mentioned previously, being the fruit of a kind of myrtle, *Eugenia Braziliensis*, and the avocado pear, an insipid fruit, generally eaten with pepper and salt, are both, to my mind, fruits which require an acquired taste in order to appreciate them. Among European fruits, the best is possibly the fig, of which there are several varieties, the most popular having a nearly black fruit. The trees, which grow mostly near the seashore, assume curiously distorted and stunted shapes, and spring from the clefts in the rocks, often overhanging the sea. They are particularly noticeable on the road between Funchal and the seaside village of Camara do Lobos. Granadillos, the fruit of different varieties of passion-flowers, some having purple fruit, others orange, suggest an exaggerated gooseberry, as the fruit when cut has much the same appearance, with large seeds embedded in a pulpy consistency. The insipid fruit of the common cactus, or prickly-pear, is much relished by the natives in hot weather, who, I was assured, gather it in the early morning, and before handling it, roll it about under their callous feet in a tub of water to get rid of the spines. The Cape gooseberry, the fruit of *Physalis Peruviana*, is prized for making preserves, and the plant has become naturalized. Many of our European fruits are cultivated, but produce fruit of a very inferior quality, the trees being seldom, if ever, pruned, and receiving little attention; but apples, pears, plums, apricots, and peaches, all come into the market in the course of the summer and autumn, while strawberries continue in bearing from the end of March till September.

PRIDE OF MADEIRA AND DAISIES

The fruit-trees are more valued for the beauty of their blossoms than their fruit by the English as a rule; and in spring, when the peach and apricot trees are laden with their pink blossoms, the country near the seashore, especially on the east side of the town, is very beautiful. The rocky nature of the ground in many places has made cultivation impossible, and stretches remain where the natural rock, covered with crustaceous lichens, appears. The shallow soil only provides a home for cactuses, which grow to an immense size; but now and then a peach-tree or a little colony of almond-trees have found sufficient soil in which to get a hold. The trees may be twisted and distorted, storm-bent by the strong winds that sweep in from the Atlantic, but for that reason are all the more picturesque; while here and there a group of stone-pines, or a group of cypresses—sentinels, guarding a little silent graveyard—give variety to the landscape, and stand out in admirable contrast to the deep blue sea below. Such plants as an occasional *Euphorbia piscatoria*, a cheiranthus, a lavender, (*Lavandula pinnata*), the Madeira stock (*Mathiola Maderensis*), some of the sedums, *Sonchus pinnatus*, of the sow-thistle family, a native of the island, and a long list of other more or less insignificant wild-flowers, may all be noticed. But by far the most beautiful and showy is the *Echium fastuosum*, pride of Madeira, which is to be seen on the cliffs along the New Road, though never with as large and perfect heads of bloom, or so deep in colour, as when cultivated. Another variety, *candicans*, has flowers of a darker blue, but is only to be found in the hills. Among this rough ground, and unfortunately in many a ravine and wall which was formerly clad with ferns and plants of a far more interesting nature, the rank-growing *Eupatorium adenophorum* seems to have taken complete possession, and threatens to become a very serious eyesore and enemy to the natural vegetation. The Portuguese have christened it *Abundancia,* and it is well named, as there seems to be no end to its abundance; its dirty-coloured fluffy heads of blossom spread their seed in all directions. It was an evil day when it was first introduced to the island as a treasure, carefully installed in a pot. Other horticultural pests have been introduced in the same way, such as the rosy purple *Oxalis venusta,* whose little flowers are pretty enough in their way, but its far-spreading roots have become a most troublesome weed in cultivated ground; and the yellow double-flowered *Oxalis cornuta* is even worse, taking complete possession in some places of any sort of grass-land. The dreaded coco, a grass growing from a tiny bulb, which throws out long and far-reaching roots, runs in the ground, till once thoroughly established, there is no end to it; this also was imported, probably accidentally, not much more than twenty years ago. The most serious of all pests in the island, the tiny black ants, the despair of house-keepers, fruit-growers and gardeners alike, were also imported from Brazil, and have gradually spread from the lower to the higher altitudes, until now I believe there is scarcely a district left in the island which is free from their ravages.

THE PURPLE BOUGAINVILLEA

CHAPTER X
CREEPERS

The year opens in Madeira with a wealth of blossom, as in the month of January the bougainvilleas, for which Madeira is so justly famous, will be in all their flaunting beauty. It is true that the lilac-coloured *Bougainvillea glabra* will have already shed most of its blossoms, as it is a summer-flowering creeper, but it is replaced by so many other varieties that its pale beauty is forgotten. The brick-red coloured *Bougainvillea spectabilis*—which must have the full force of the sun upon it in order to bring out its colour to the best advantage, being apt otherwise to look a false colour—when grown over pergolas, or corridors as they are called in Madeira, or allowed to wander at will over a wall or bank, provides a gorgeous mass of colour. I had seen bougainvilleas in other countries, but only grown against walls, and closely cropped by shears, in order that the wood might be sufficiently ripened by the heat of the summer to insure its wealth of blossoms. Here such care is not necessary, and the natural beauty of the plant can be seen to full advantage where it has escaped the ruthless shears of the Portuguese gardener. Branches of blossom, ten, fifteen, or even twenty feet long, show the strength with which the plant grows; in fact, many a splendid specimen has had to be sacrificed, for fear it should undermine a terrace wall or shake the very foundations of a house.

To the landscape gardener who is fastidious as to the scheme of colouring in his garden, the placing of all the varieties of bougainvillea (called after the French navigator, De Bougainville) forms one of his chief difficulties. Each in itself seems too beautiful to be discarded; but, unless the garden is of considerable extent, I would recommend the owner of the garden to harden his heart and make his choice of the colour he prefers and stick to it, only growing the one variety in some great mass, be it as the gorgeous canopy of his corridor, or clothing his garden-wall.

Many persons give the palm for beauty to the deep magenta variety, *speciosa*, as it stands alone for colour. In all the kingdom of flowers I know no other blossom of the same tone of colour; it is a thing apart, this royal purple flower. No one who has seen the plant which covers the cliff below the fort can ever forget its beauty. Seen from the sea, it stands out like a purple rock in the middle of the city. By the middle of January it will be in all its gaudy, garish splendour, the admired of all beholders.

It can well be imagined how these two varieties—the one brick-red, the other deep magenta—would strike a jarring note in any garden if grown side by side, or even within sight of each other. And do not imagine that Madeira only boasts of these two coloured bougainvilleas in its winter season. From these two have sprung many others—seedlings, no doubt, hybridized in a country where the heat of the sun will ripen most seeds. So now there are rosy reds, lighter or darker, to choose from, shading through a range of colour which, like the beauty of its parents, seems to stand alone.

The plant has, I consider, two enemies in the island. One is the ordinary uneducated Portuguese gardener, who seems to think that the art of gardening consists in so closely pruning a creeper or shrub that all the natural grace and beauty of the plant is lost for ever, as often as not choosing the moment for this cruel treatment when the plant is in full flower. Though Nature has done her best to protect the plant from the hand of man, by giving it long, hooked thorns, which are exceedingly sharp, and, I believe, somewhat poisonous, even this has not been sufficient, and many a beautiful specimen have I seen maimed and dwarfed beyond repair in a few hours by an ignorant and overzealous gardener. Its second enemy is rats, which unfortunately have a great love for the bark on the stems of old plants, and many a plant narrowly escapes destruction at their hands, or rather teeth.

The second place in the list of creepers for the New Year must be given to the flaming orange *Bignonia venusta*, a native of South America, with its dense clusters of finger-shaped flowers. This has now become the commonest of all creepers in Madeira, and there is hardly a road in the neighbourhood of Funchal where all through the month of January there is not a stretch of wall bearing its gaudy burden, or a *mirante* (as the arbour or summer-house dear to the hearts of the Portuguese is called) without its roof of golden blossoms. There is a long list of bignonias and tecomas—a family so closely allied to each other as to be almost united—whose full beauty is for a later season; and only stray blossoms of the deep red *Bignonia cherare*, with its long yellow-throated trumpets, appear in the winter months, but sufficient to give promise of glories to come in the month of April.

BIGNONIA VENUSTA

In the same month the close-growing *Tecoma flava* will become wreathed with its golden-yellow trumpet flowers, clothing many a wall and straying across tiled roofs, as it is so neat and clinging in its habit that it never becomes so heavy a mass as to damage buildings. Its companion at the same season is *Tecoma Lindleyana,* bearing large mauve trumpet flowers, with a throat of a lighter shade. The individual flowers are of extremely delicate texture, and are beautifully veined with a slightly darker shade of purple. Yet another tecoma unfurls its blossoms late in the month of April, but is not so often met with as the two former varieties, possibly because the plant, when out of flower, presents rather an unsightly and straggling appearance; but no one can fail to admire the pure white and yellow-throated blossoms of this *Tecoma Micheliensis,* as it is most commonly called, though I believe it has a second, and possibly more correct, name.

For May and June is reserved, probably, the most beautiful of all the tecomas, *jasminoides.* The plant is an ornament at all seasons; its beautiful glabrous foliage seems to retain its freshness at all seasons of the year, and when the plant is covered with its bunches of large white blossoms, each with its deep red-purple throat, which seems to reflect a shade of purple on to the white petals, it is one of the most beautiful of all creepers.

From the list of winter creepers the *Thunbergia laurifolia,* with its bunches of grey-blue gloxinia-shaped blossoms, cannot be omitted; though the beauty of the plant is somewhat spoilt by the habit of the dead blossoms hanging on instead of falling, and marring, by their brown, shrivelled appearance, all the freshness of the newly developed flowers. The plant always recalls to my mind the reason given by the Japanese for not admiring the national flower of England—the rose—as they complain that it clings with ungraceful tenacity to life, as though loath or afraid to die, preferring to rot on its stem rather than drop untimely; unlike the blossoms of spring, ever ready to depart life at the call of Nature. Such is certainly the case with thunbergia. The creeper is also a dangerous poacher, and, unless kept within bounds, will soon smother and overwhelm any shrub or tree that it may take possession of, though never in Madeira attaining to the vast proportions that it assumes in Ceylon or other tropical countries, where it takes possession of even the tallest forest-trees, and hangs its long trailers from one tree to another, and on and on again, in one dense tangle. The white variety does not seem to have been introduced to Madeira, and its pure white blossoms recall gardens in St. Vincent and other West Indian islands.

Yet another creeper whose flowering season belongs to the winter months is the scarlet passionflower, *Passiflora coccinea*. By the end of January the plant will be covered with a few fully opened flowers, many half-developed flowers and innumerable buds giving promise of its future splendour. On first acquaintance, one is deceived into thinking that in a few days' time the plant will be a sheet of scarlet blossoms, but such is not the case: each individual flower is short-lived, and by the time the half-developed blossoms have opened, the fully expanded blooms of yesterday have vanished. Thus its flowering season is a prolonged one, but it never attains to any very gaudy splendour.

By the last days of March the racemes of that most beautiful of all creepers, *Wistaria chinensis*, or *sinensis*, will have begun to lengthen, and gradually clothe the whole plant with a pale purple canopy. The vine—as it is called the grape-flower vine, from the resemblance of its blossoms to a bunch of grapes—is a native of China and Japan, and also of parts of North America, which accounts for the fact that it received the name of wistaria (by which it is known all over the Western world) from one Caspar Wistar, a medical professor in the University of Pennsylvania. In Japan the plant is known as *fuji*, and is so universally admired that, in common with many other flowers, it is made the excuse for many a flower-feast, when hundreds, thousands, and even tens of thousands of pleasure-seekers will hold their revels, or sit quietly sipping their tea under a roof of the royal *fuji*. Though in Madeira it is not the fashion of the country to hold flower-feasts, or to make flowers the theme of poems and plays, or to regard wistaria as an emblem of gentleness and obedience, as is the case in its Eastern home, yet in this land of its adoption it comes in for its full share of admiration. Corridors and walls which have been passed by unnoticed through the winter months, having been only clad with the long, bare, leafless branches, the last leaves having fallen early in December, suddenly become transformed, and for a few short days—all too short, alas!—become the centre of attraction in the garden.

Like in Japan, the wistaria season begins with the white wistaria, which has been christened in the Western world *Wistaria Japonica*, and "it would seem as though this modest white wistaria had been allowed by Nature to bloom so early, for fear she should be overlooked and not appreciated when her more showy successor flings her purple mantle over the land." There are good specimens of this early white variety in the gardens of the Quinta da Levada and the Quinta do Val.

The variety known as *Wistaria multijuga*, for which Japan is so justly famous, as it appears to be the only country where its full beauty can be seen, has been introduced with but little success to the island. It is true that it will grow, and grow strong, but its long racemes of thin, pale, washed-out-looking flowers are but a sorry sight to those who have ever seen the far-famed Kameido Temple grounds in Tokyo, when the vines, with their long purple tassels, often over three feet in length, clothe the long trellises and almost smother the guests who sit feasting beneath them, gazing across at the long vista of mauve blossoms reflected in the water below. But even in Japan this far-famed *multijuga* variety is only to be met with in certain districts and as a cultivated form, and is never seen clambering from tree to tree in a wild state, like the *chinensis* variety. The wistaria season closes as well as opens with a white-flowered form both in Japan and Madeira, as the variety known as *macrobotrys*, with its very long racemes of white blossoms, prolongs the beauty of the *fuji* feast at the celebrated Kameido Temple; and here in Madeira, though only one or two plants of it exist, it is the last to retain its beauty.

The summer months will have their own creepers, though not such showy ones as the winter and spring months; but if they are lacking in colour, many of them atone for that by their delicious fragrance. To these belong *Rhyncospermum jasminoides*, or *Trachelospermum*, as I believe it is more correctly called, whose white starry flowers fill the whole air with their almost overpowering scent. The plant is a native of China and Japan, where it may be seen growing in a perfectly wild state in hedgerows. There is another variety called *angustifolium* whose blossoms are much the same, but the foliage differs, and this kind is said to prove hardy when grown against a wall in the South of England. The well-known *Stephanotis floribunda*, called in its native country the Madagascar chaplet flower, unfurls its heavy-scented waxy blossoms in the summer months. *Allamanda schotii*, hoyas, with their clusters of waxy red blossoms, mandevilleas, and hosts of others, are seldom seen in their beauty by the English owners of gardens.

CHAPTER XI
TREES AND SHRUBS

The list of indigenous and naturalized trees and shrubs growing in Madeira is such a long and varied one that it is not surprising that Captain Cook, in his account of his first voyage, should have said: "Nature has been so liberal in her gifts to Madeira. The soil is so rich, and there is such a variety of climate, that there is scarcely any article, either of the necessaries or luxuries of life, which could not be cultivated there."

The place of honour among the island trees must be given to those belonging to the laurel tribe, of which there are a great number, and splendid specimens still remain in the country, survivors of the wholesale destruction of the primeval forests. To this tribe belongs the til, one of the most beautiful of evergreen trees, its shiny green leaves contrasting admirably with the light grey bark of its stems. The old trees grow to a very large size, and in the Boa Ventura Valley and along the road to Sao Vincente there remain some grand old specimens, the immense girth of whose trunks speaks for itself of their great age. The true name of this so-called laurel appears to have been a matter of some uncertainty, as Miss Taylor, in "Madeira: Its Scenery, and How to See It," classes it as *Oreodaphne fœtens*, describing it as "the grandest of native trees"; while Mr. Bowdick, in 1823, says: "The til has been confounded with *Laurus fœtens*, from the strong, disagreeable odour of the wood when first cut. It is very valuable for its timber, being extremely hard and tough. It would appear that the Portuguese call both *Laurus fœtens* and *Laurus cupuleris* til, as they say there are two kinds of til, and both are equally fetid." In the damper regions beautiful lichens grow luxuriantly on the stems of the trees, and ferns have found a home in the cracks of the bark. The value of its timber has no doubt been responsible for the destruction of the trees. When polished, the wood is of a very dark colour, almost as black as ebony.

The vinhatico, whose wood is the mahogany of Madeira and closely resembles it, is another of the native trees, and again I find it classed as *Laurus indica* by Mr. Bowdick, who describes it as one of the island's most valuable products, while Miss Taylor describes it as *Persea indica*. The wood, when cut, is of a deep red colour before being polished. It is a fine forest tree, and has, as a rule, light green foliage, though it occasionally turns crimson. It has given its name to one of the most beautiful bits of scenery in the island, as the Levada dos Vinhaticos, running above the village of Santa Anna, passes through some of the grandest scenery in Madeira. Professor Piazzi Smyth has gone so far as to assert, in "Madeira Spectroscopic," that some of the largest ships of the Spanish Armada were either built of, or internally decorated with, the wood of the tils and vinhaticos of Madeira. This would appear to be a flight of imagination, or a revelation of the learned man's inner consciousness, as it is difficult, if not impossible, to find any grounds for such an assertion, there being no document extant stating what timber was employed for the building of that celebrated fleet.

The laurel familiar to us under the name of Portugal laurel, *Cerasus lusitanica*, assumes the proportions of forest trees, and when I saw it in spring, covered with its long racemes of creamy-white flowers, it quickly dispelled the aversion with which I had always regarded the stumpy, blackened specimens pining under the smoky atmosphere of suburban shrubberies.

Laurus Canariensis is a fragrant form of laurel, and the country-people extract oil from its yellow berries.

Picconia excelsa, the *Pao branco* of the Portuguese, is generally to be found in the same districts as the til-trees, and attains to a height of forty or fifty feet. Its hard, heavy white wood, being in great demand for the keels of boats, is very valuable. Like many other native trees, it is for this very reason rapidly becoming scarce, as its destroyers, having no thought for the future, omit to cultivate it from seed, which grows readily.

The *Clethra arborea*, or lily of the valley tree, as it is called by the English, on account of the resemblance of its spikes of creamy-white flowers to those of a lily of the valley, fills the whole air with its delicious though somewhat heavy fragrance when the tree is in flower in summer. Yet another fragrant tree peculiar to Madeira is the *Pittosporum coriaceum*, which has been christened the incense-tree, as early in April the air, especially near sundown, is filled with the almost overpowering scent of its clusters of small greenish-white flowers. The bark is very smooth and even, and of a light ash colour. The tree is now somewhat rare in its natural state, but is frequently seen in gardens, where it has no doubt been transplanted from its original home among the rocks, as Mr. Lowe, in his "Flora of Madeira," remarks how he only noticed it growing on high rocks or in inaccessible places.

One of the first trees which is sure to strike the eye of the new-comer is the dragon-tree, or *Dracæna draco*, on account of its peculiar growth. From having been a common tree on the island it has now become a rare one in its native state; in fact, the only ones I have ever seen under those conditions are a few sole survivors on the rocks beyond the Brazen Head, where formerly they grew in great numbers. Now by their quaint growth they give a distinctive feature to many a garden, and it is consoling to know that they are easily raised from seed. Mr. Bowdick, in writing of the tree, says: "The dragon-tree was considered by Humboldt as exclusively indigenous to India, but I am inclined to think it is also natural to Porto Santo, and perhaps to Madeira—not from the few specimens which now remain on these islands, but from the account of Cadamosto, who visited Porto Santo in 1445, and writes that the dragon-trees of Porto Santo were so large that fishing-boats capable of containing six or seven men were made out of the trunks, and that the inhabitants fattened their pigs on the fruit; but he adds that so many boats, shields, and corn-measures had been made out of them, that even in his time there was scarcely a dragon-tree to be seen in the island."

The stem exudes a gum, and the following account of the means of collecting it is taken from a Portuguese account of "The Discovery of Madeira," written in 1750: "All over the island grows a tree from which the dragon's blood is procured. This is performed by making incisions in the bark, from whence the gum issues very plentifully into pots hung upon the branches to receive it. The people use it as a sovereign remedy for bruises, to which they are very much exposed by traversing their rocky country; and this, with one panacea more, completes their whole *materia medica*—that is, balsam of Peru, imported from the Brazils in small gourds by their annual ships. These two, they imagine, have power to cure almost all disorders, especially those that are external."

Among other native trees, the beautiful *Taxus baccata* and the *Juniperus oxycedrus*, with its great spreading silvery-green branches, cannot be omitted. The former has become almost extinct, and the juniper is also becoming rare, from the reckless way in which the trees have been cut to be used for torches. The fragrant red wood is split into lengths, and several bound together, for this purpose. In gardens their dense growth makes them admirably suited to form an arbour, in the absence of the ubiquitous *mirante*, as they provide shelter from the wind and perfect shade.

Another evergreen tree, which, though not a native tree, is very commonly to be seen in and about the town of Funchal, is the *Ficus comosa*, which, as its name implies, is a beautiful tree, though, from its having such far-spreading hungry roots, it is more suited to the roadside than to gardens. A peculiarity of the tree is the slenderness of its stem in comparison to the immense length and weight of its very spreading branches; its bark is a very light grey colour, and is in admirable contrast to the very smooth and shining leaves, which are dark green above and pale beneath, produced in masses on the slender rather hanging branchlets. Two very fine specimens of these trees stand alone on the Rodondo, near the Quinta das Cruzes, from where a very fine view of the town is to be seen from under their immense spreading branches.

JACKARANDA-TREE.

The camphor-trees are at their best in spring, when they are covered with their delicate young green shoots, generally of a very light green, but occasionally having brilliant red shoots. The trees attain to a large size, though not assuming the gigantic proportions which they reach in their native land, Japan. That most uninteresting of all trees—the plane-tree—has been planted along the beds of the rivers in the town; and the oaks are in almost perpetual foliage, as the young leaves appear before the old ones have really fallen.

Grevillea robusta is common in gardens, where, having shed its leaves in winter, the trees are showy in the early summer months, being covered with yellow flowers; but the palm for flowering trees must be given to the *Jacaranda mimosafolia*, a native of Brazil. Having also shed its long fern-like foliage in the late winter months, early in May the tree bursts into a cloud of blue blossoms, almost as blue as the sky above. The tree is a fairly common one in and about Funchal, and the "blue trees," as they are generally called, are the admired of all beholders during the few weeks they are in bloom. Nature has done well in ordaining that the foliage should fall before the tree blossoms, as the full beauty of the flower is thus seen unshrouded by leaves.

The list of flowering trees is a long one, but I cannot help mentioning a few others which are ornaments to the gardens when in bloom. The dark red of the *Schotia speciosa* blossoms also adorn a leafless tree. The tree, which was called after a Dutchman, one Richard van der Schot, in its native country — subtropical Africa — is commonly known as the Kaffir bean-tree, no doubt because its blossoms are more suggestive of bunches of red seeds or beans than flowers.

There are a few specimens of the gorgeous *Poinciana regia*, which flowers in summer; its peculiar flat, spreading branches are easily recognized. No one who has ever seen these magnificent trees in all their gaudy splendour in tropical regions can ever forget their beauty. They deserve their name, the royal peacock flower, though they are more commonly known as flamboyant-trees, from the likeness of their leafless branches, clad with brilliant orange-red nasturtium-like blossoms, to flaming torches. In Madeira the tree does not attain to its full beauty, as possibly the difference between the climate of its native home — Madagascar — and that of Madeira is too great. Here the less showy variety, known as *Poinciana pulcherrima*, thrives better.

At the same season the uncouth growth of the bare and leafless frangipani or plumeria trees bursts into blossom — white, cream-coloured, or pale pink — and fills the air with its heavy fragrance, recalling the oppressive, almost stifling, atmosphere of Buddhist temples in Ceylon, where frangipani blossoms are almost regarded as sacred to Buddha, and are always called "temple flowers."

Of the coral-trees there are several varieties: *Erythrina corallodendron*, a native of the West Indies, has large spikes of deep red blossoms on leafless light grey stems; and *Erythrina cristagalli*, a native of Brazil, also bears scarlet blossoms. Besides the flowering trees, there are so many shrubs which contribute such a wealth of colour to the gardens, especially in the winter months, that it is hard to decide which are most worthy of notice. The gaudy orange-coloured *Streptosolen Jamesonii*, which was only introduced into Madeira a comparatively short time ago, has now become one of the commonest, but none the less beautiful, of winter-flowering shrubs. Like many other plants which I had only known pining in the unfavourable atmosphere of an English greenhouse, it is almost impossible to recognize the streptosolen of the greenhouse, with its dull orange and yellow blossoms, as the same plant when grown in the sunshine of Madeira. The soil is no doubt partly responsible for the difference in colour — a fact I have noticed with many other plants, but certainly in the case of streptosolen the change is most remarkable—and the intense brilliancy of its large heads of blossom attract the attention of all new-comers to the island. The shrub is sometimes known as *Browallia Jamesonii*; and a blue variety which has lately been introduced from the Cape seemed to closely resemble the family of browallias. Should it prove to have as vigorous a constitution as the orange variety, it will be another great acquisition to the island, as its blossoms are of a deep clear blue.

Astrapæa pendiflora, or tassel-tree, as it is often called, from the resemblance of its great balls of pink blossoms hanging on a long slender stalk, has handsome foliage, and assumes the proportions of a large shrub or small tree in a short time, as it appears to be of very rapid growth. I find it difficult to share the almost universal admiration that it awakens when in flower, as its beauty is much marred by the tenacious habit of its dead blossoms, which cling to life to the bitter end, and spoil all the freshness of the newly developed blossoms. The balls of blossom, in shape reminding one of huge guelder roses, start by being a greenish-white, which gradually turns to a deep dull pink, and in death to a most unsightly brown. *Astrapæa viscaria* attains to the size of a large tree, and in April bears a burden of pink blossoms, also in round balls; it is a native of Madagascar, which seems to be the home of so many of the most beautiful flowering trees.

Among purple flowering shrubs, for the beauty of its individual flowers and purity of colour, *Lasiandra* or *Pleroma macrantha*, with its large deep violet-purple blossoms, deserves a place in every garden. The plant cannot be reckoned amongst the most showy of the flowering shrubs, as it does not bear many blossoms fully expanded at the same time, though, as the flowers are very freely produced at the ends of the branchlets, its flowering season is a prolonged one. The plant appears to be a native of Brazil, which is another home of many of the most beautiful of flowering shrubs.

Wigandia macrophylla attains to the size of a small tree; its large, loose heads of lilac-purple flowers, somewhat resembling paulonia blossoms, and its handsome foliage, combine to make it a most ornamental plant and a valuable acquisition all through the winter and early spring. To Brazil we owe another favourite shrub, *Franciscea latifolia*, as it is commonly called, though it appears to belong to the *Brunsfelsias*, a family of shrubs called after one Otto Brunsfels, who was first a Carthusian monk and afterwards a physician. The clear lilac blossoms have a distinct whitish eye, and as they fade, turn to a greyish-white, so the shrub appears to bear white and lilac blossoms at the same time. The blossoms are deliciously fragrant, though many people consider their scent to be too strong and overpowering. A well-grown specimen attains to eight or ten feet, and has pleasing shiny green foliage.

The light crimson-flowered *Hibiscus rosa sinensis*, which ornaments most gardens in tropical or subtropical regions, has also found a home in Madeira, and the long white trumpet-flowering *Brugmansia suaveolens*, more commonly called daturas, natives of Mexico, have found so congenial a home that the shrub may almost be considered to have become naturalized. Growing at the bottom of many a ravine rich in vegetation, the shrub will appear to be in a perfectly wild state, bearing a fresh crop of leaves and blossoms with every new moon, and filling the air at nightfall with their heavy scent.

The blossoms of the daturas are known as *bellas noites* by the Portuguese, though the night-scented flowers of *Cestrum vespertinum* seem to share the name with them; occasionally, it is true, the latter are deemed masculine, and are therefore called *boas noites*. The following interesting description of *Brugmansia* or *Datura suaveolens* is taken from Mr. Lowe's "Flora of Madeira," written in 1857: "The flowers are slightly fragrant by day, but much more powerfully and diffusedly so after sunset and through the night, when, by moonlight, they display an almost radiant or phosphorescent snowy-whiteness, and expand more fully, falling into elegant thick horizontal rows or flounces on the trees or bushes. Nothing can exceed their grace and loveliness when in full luxuriance and perfection, which it may be said to attain at intervals of four to five weeks continuously, from June to November or December. The tree is esteemed noxious, and therefore in Madeira of late years has been banished from gardens and near proximity to houses. This idea perhaps originated from an accident which occurred some forty years ago, when two or three children, having eaten a few of the seeds, escaped by timely medical assistance, with no further harm than the effects of an overdose of *Atropa belladonna*. Still, there is something perceptively oppressive in the evening, in too long or close inhalement of the powerful aromatic fragrance of the flower."

The peculiar flowers of *Strelitzia regina*, introduced to Europe from South Africa during the reign of George III., and named, in honour of Queen Charlotte of Mecklenburgh Strelitz, never fail to attract admiration. The plant is also called bird of paradise flower and bird's-tongue flower—both suitable names, as the gaudiness of its blue and orange flowers must have been responsible for the former, while the resemblance of the flower to a bird's head with a bright blue beak shows its likeness to the latter. The plant has long, narrow, oblong leaves, of a dull greyish-green, of a peculiarly tough texture, and a good clump some four or five feet high is very ornamental.

Strelitzia augusta, as its name implies, is of more majestic growth. It has large foliage, not unlike a banana, and clumps attain to twelve or fifteen feet in height. The blossom is more curious than beautiful, being of so dark a purple as to be almost black; but, for the sake of its foliage, it is always worth a place, and may well be called a noble plant.

A CHAPEL DOORWAY

CHAPTER XII
HISTORICAL SKETCH

Though this volume does not profess to be in any sense a guide-book to the Island of Madeira, yet it seems as though even those visitors to the island, who may only wish to study its flora and sylva, will more fully appreciate their wanderings by learning something of its history.

Very little is known of the early history of Madeira. Though some historians assert that even the early Phœnicians found their way there during some of their adventurous voyages, there seems to be little foundation for such assertions. Others at a later date claim for Madeira the honour of being Pliny's Purpuria, or Purple Land, an honour to which the Canaries also lay claim, though it seems probable that Madeira has more right to the distinction, as Humboldt gave new life to the theory by describing in glowing terms the beauties of its hazy mountains, shrouded in purple and violet clouds. A less romantic reason for the name of Purple Land is also given, and merely relates to the fact that King Juba in the days of Pliny contemplated the possibility of extracting a purple dye, called "gætulian purple," from the juice of one of the numerous trees or plants which grew on the island. This theory is supported by its upholders by the fact that Ptolemy mentions an island in this part of the Atlantic Ocean called Erythea, or Red Island, which again may possibly have reference to the dye. After these early days there is no trace of the island in history for hundreds of years, so it is more than problematical as to whether the Purple Lands had any connection with Madeira.

There seems to be no end to the number of legends and vague theories as to the discovery of the group of islands. An Arab historian relates the discovery of an island (possibly Madeira) by an expedition of his people in the eleventh century, who gave it the name of El Ghanam. These travellers, known as the "Almagrarin adventurers," set sail from Lisbon with the intention of discovering something. Their name, meaning the "finders of mares' nests," is suggestive of fabulous tales. After being driven across unknown seas they came to a district of "stinking and turbid waters," which at first frightened them back; and it is suggested that, as the soil of Madeira shows traces of volcanic disturbances—as, indeed, does the whole formation of the island—these disturbed waters might well have been in its neighbourhood.

In the fourteenth century both the French and Spaniards claim to have touched at the islands; but if such were the case, it seems unlikely that their discovery would have been relegated to oblivion, though in the Medici map in Florence the group of islands now known as Porto Santo, Madeira, and the Desertas appear, under the names of "Porto Sto," "Ila Legname," and "I. Deserta." If these names were inserted when the chart was made (a.d. 1351), the Genoese might claim to have been the true discoverers; but as the names are merely Italian translations of the Portuguese, it is more likely that they were added after their present owners had taken possession of them.

It is through the medium of another legend, as some still call the romantic story of Machim and his lady-love, Anna Arget, or Harbord, that we appear to arrive at the true history of the discovery of Madeira. The story, though it is more suggestive of fabulous romance than history, has been accepted as being the medium of the tales of the unsurpassed beauty of the island coming to the ears of the enterprising Portuguese navigator Joao Gonsalvez Zargo. The tale relates how one Robert à Machin, in the reign of Edward III., fell in love with a beautiful young lady of noble family named Anna d'Arget. Being endowed with great wealth as well as beauty, her parents destined her for a greater match, which was accordingly arranged. Though the lady returned her young lover's affection, she was compelled, in an age when the daughters of a great house had little voice in the choice of their husbands, to marry the nobleman chosen by her parents. In order to insure that their plans should not be frustrated, the lady's parents went so far as to arrange that her lover Robert should be imprisoned until after the marriage. When he was liberated he heard from a friend of the fate of his lady-love, and lost no time in following her to her new home and arranging for their elopement. This took place by sea, the adventurous couple embarking at Bristol, hoping to make the coast of France. Contrary winds arose, and we are told that, after enduring great perils and hardships for thirteen days, Robert and Anna, accompanied by a few faithful followers, came to "a pleasant but uninhabited land, diversified by hills and vales, intersected by clear rivulets, and shaded with pine-trees."

Dr. Gaspar Fructuoso, in his work entitled "As Saudades da Terra," written in 1590, tells of the lovers' great joy when, "on the morning of the fourteenth day, when they had been hourly expecting destruction, and were in a hopeless and exhausted condition, they saw a dark object before them, which they imagined might be land, and when the sun rose they perceived that their surmises were correct and their hopes fulfilled. As they drew near, they saw that the mountains rose, as it were, almost directly from the water's edge in many places. The almost perpendicular cliffs seemed to preclude any landing, except where the grand ravines opened right down to the sea. It was into one of these openings of enchanting loveliness that Machim directed his vessel to be steered, and, casting anchor, a boat was most eagerly launched. Machim and some companions hurried on shore, and they soon returned with such an encouraging account that he took his beloved Anna from off the vessel where such terrible and anxious days had been passed, and landed on a shore where he hoped he should, with such comforts as still remained to him, procure for her, for a time at least, some repose, refreshment, and security."

For some time the party devoted their time to exploring their immediate surroundings, in a land which appeared to them a haven of rest and of surpassing loveliness. They penetrated into forests of great extent, to points on the mountain-tops from whence a succession of wooded ravines and steep mountain-sides, clothed with a luxuriant and ever-verdant vegetation, delighted their eyes; the mountain streams giving life to a scene where, except only for the songs of countless birds and the hum of insect life, all was still. No four-legged animals or reptiles were to be seen. Fruits in abundance seemed as if awaiting them, and in the crannies of the rocks they found honey possessing the odour of violets. An opening in the extensive woods, which was encircled by laurels and flowering shrubs, presented an inviting retreat, and a tree of dense shade, the probable growth of ages, offered a verdant canopy of impenetrable foliage. In this spot they determined to form a residence from the abundant materials with which Nature supplied them. This state of innocent happiness was not destined to last long, as, though apparently serenely contented with their surroundings as long as the vessel anchored close at hand suggested a possible retreat and return to the outer world, disaster befell them, for one night a storm arose and their ship was driven out to sea. This calamity so greatly distressed the fair lady that she became completely prostrated by the shock, and in a few days she died in her lover's arms. Machim, in his turn, died of grief a few days after, having spent the intervening time in erecting a memorial to his much-loved Anna. The dying man dictated an inscription recording their sad story, concluding with a request that if any Christians should at any future time form a settlement in that island, they would erect a church over their graves and dedicate it to the Redeemer of Mankind, a request which, it will be seen, was afterwards carried out, when "Machim's tree" was supposed to have furnished sufficient material for the building of the whole chapel.

Their survivors not unnaturally set about building a boat in which to escape from the land which by now was filled with sad associations for them, and eventually they succeeded in reaching the coast of Morocco. Here a worse fate awaited them, as they fell into the hands of the Moors and became slaves. They are said to have joined some of their fellow-comrades who had been on the ship when she was driven out to sea. Their past and present adventures, and the descriptions they gave of the beauty of this fairy island, attracted the attention of a fellow-slave, a Spaniard named Juan Morales, an experienced pilot.

Morales treasured all this information, and was eventually ransomed through the intervention of his Sovereign. On his return to Spain he was taken prisoner by the Portuguese, and carried off to Lisbon by Joao Gonsalvez Zargo, the celebrated navigator, who lost no time in informing his patron Prince Henry of the tales he had heard from his prisoner of the fertility and beauty of the undiscovered island.

Prince Henry was the son of John I. of Portugal, and a nephew of our Henry IV. He was called "O Conquisador," and the Portuguese are justly proud of him, as through his love of exploration and adventure he added largely to their dominions, and lent a ready ear to rumours of undiscovered lands. Zargo had no difficulty in persuading his patron to fit out an expedition, which he himself was appointed to command. On June 1, 1419, he set sail for Porto Santo, which had been discovered two years previously by the Portuguese. The colonists on the island related how, in one particular direction, there hung perpetually over the sea a thick, impenetrable darkness, which was guarded by a strange noise which occasionally made itself heard. With the usual superstition of the age, various reasons were ascribed to these mysterious signs. We are told "by some the place was deemed an abyss, from which whoso ventured thither would never return; by others it was called the Mouth of Hell. Certain persons declared it to be that ancient island Cipango, kept by Providence under a mysterious veil, where resided the Spanish and Portuguese Christians who had escaped from the slavery of the Moors and Saracens. It was considered, however, a great crime to dive into the secret, since it had pleased God to signify His intention to reveal it by any of the signs which were mentioned by the ancient prophets who spoke of this marvel."

Being less superstitious and more adventurous than these benighted colonists, Zargo determined to fathom the mystery of this so-called impenetrable darkness. Setting sail one morning with a fair wind, by noon his hopes were fully realized, and he found the mysterious veil to be nothing more than heavy clouds hanging over the densely wooded mountains on the north of the island—a state of things very commonly seen to this day when approaching the island from the north. Like the unfortunate couple Machim and Anna, he was filled with joy and delight when he saw the grand mass of mountains rising abruptly from the sea. The party soon found themselves sailing along a glorious coast, with grand cliffs, cut by deep densely wooded ravines, coming down to the sea.

On the morning of June 14, 1419, having anchored for the night in a sheltered bay, which exactly corresponded with the description given by Morales, who accompanied the expedition, of Machim and Anna's resting-place, Zargo and some of his followers landed—and this is the first authentic account of the discovery of Madeira.

The party spent some days exploring this rich and fertile acquisition to the Crown of Portugal, and on July 2 Zargo, accompanied by two priests who formed part of the expedition, held a ceremonious service of thanksgiving for the discovery of the island, taking formal possession of it in the name of the King of Portugal. Mass was celebrated and a service was held on the spot which was supposed to be the grave of the two lovers. The final ceremony consisted in the laying of the foundation-stone of a chapel dedicated, in accordance with Machim's request, to the Redeemer of Mankind.

Before returning to Portugal to announce the joyful news of his discovery, Zargo explored the coast, and named various points and bays with the names they still bear at the present day. Machim's bay was named Machico, and may claim to be the oldest settlement. The most eastern point of the island had already been named Ponta de Sao Lourenso when the travellers rounded it—some say because Zargo, calling for the aid of St. Lourenso, after whom his ship was named, jumped into the sea at this point and landed; others assert that the point was merely named after one of his companions who bore the saint's names.

Santa Cruz was so named because at this spot the party found some large trees lying on the shore, torn up by the elements, out of which they formed a large wooden cross. Porto do Seixo owes its name to the freshness and purity of its spring water, for which it is still famous; and the explorers were so struck by the great springs of pure water which gush out of a grand mass of rock, that they took back with them to Portugal a bottle of the water as an offering to Prince Henry.

Rounding a prominent headland which was then clothed with numerous dragon-trees, and remained famous for them for many hundreds of years, though now only one or two of the trees are left, flocks of tern were startled from their resting-place by the strange and unknown noise of oars, and flew all round the boats, even alighting on their occupants. The headland therefore received the name of Capo do Garajao, or Cape of the Tern, though at the present time it is better known to the English under the name of the Brazen Head.

From this point they saw a fine expanse of country, and at once settled that this would be the best spot on which to build the future city. As the district was remarkable for the thick growth of fennel, which in Portuguese is called *funcho*, the site of the new town received the name of Funchal.

Ribeiro des Soccoridos (river of the rescued) was the name given to a place where two of the party lost their footing whilst attempting to cross a river, and would have been swept into the sea if their companions had not come to their rescue. Praya Formoso was aptly named "beautiful shore." The extent of their wanderings on this occasion seems to have led them to the great cliff which towers some 2,000 feet above the sea, so they named the cape Cabo Girao. Having been startled by seeing some seals leaping out of caves in a bay before they approached the great cliff, they named the spot Camara do Lobos, or Wolves' Lair, which is the site of the picturesque village which was afterwards built in the sheltered situation.

From this time the history of the island is no longer wrapt in mythical legends, and it seems certain that in the following year (1419) Zargo and one Tristao Teixeira were permitted to return. They divided the island into two *comarcas*, each taking command of one: Zargo became the *Capitao*, and Teizeira the *Donatorio*, and they portioned out the land among their followers. Zargo founded the town of Funchal, and the two Captains had complete jurisdiction granted to them by the Crown, though they had to appeal to their monarch in cases of life and death. Zargo lived to enjoy his command for forty-seven years, and his tomb is still to be seen in the church of the Convent of Sta. Clara, which was founded by his granddaughter, Donna Constanca de Norouka, in 1492. Fructuoso gives an account of some of the first inhabitants of the island, and tells us that the first children who were born in the island were the son and daughter of Gonzalo Ayres Fereira, one of Zargo's companions, and they were christened Adam and Eve. Adam, the first man, founded the Church of Nossa Senhora at the Mount.

The wife of Christopher Columbus being the daughter of Perestrello, the Governor of the neighbouring island of Porto Santo, possibly led to Christopher Columbus visiting Madeira. The house which he was said to have occupied during these visits, the property of Jean d'Esmenault, was ruthlessly destroyed in the year 1877 to make room for new shops. The American Consul of that date, evidently sharing the love of the rest of his country-people for souvenirs, carried away to America many of the architectural treasures of the house, such as the carved window-frames and ornamental stonework. Thus Funchal lost one of her most interesting relics of the past.

In the year 1566 Funchal suffered at the hands of a French naval expedition which had been fitted out by Peyrot de Montluc, son of the Marshal, for the purposes of exploring unknown lands and seas, according to the spirit of adventure which was the fashion of that age. Meeting with storms, which probably diminished the number of his crew, Montluc put into Madeira, with the intention, it is said, of recruiting his force; but being eyed with suspicion, as belonging to the navy of a foreign country, he professed to have been insulted, and attacked the town. The city appears to have been feebly defended, although Montluc must have met with some resistance, as over 200 of the inhabitants lost their lives. Very little is known as to the strength of the invading force, but it is certain that great damage was done to the town by the Huguenot invaders, as they were, of course, described by the Catholics. The churches seem to have suffered severely, as the plunderers no doubt expected to find treasure in their vaults. Having thoroughly ransacked the town and terrified the inhabitants, who mostly fled to the country, the expedition departed before assistance came from Lisbon, but not before the leader Montluc had been mortally wounded. In 1580 the island, being a Portuguese possession, fell with its mother-country under the rule of Spain—a state of affairs which lasted some eighty years. Madeira seems to have been little affected by the Spanish yoke, the most important alteration in its government being the abolition of the office of Captains and the appointment of a Governor of the island—an office which the Portuguese confirmed when it again came under their sole power, and is continued to this day.

The eighteenth century appears to have been a more peaceful epoch in the history of the island, though it is recorded that Captain Cook, when starting on his voyage round the world in the *Endeavour*, bombarded the fort on the Loo Rock as a protest against an affront which he said had been offered to the British flag.

During the seventeenth century many English families settled in Madeira, as, in consequence of the marriage of Charles II. with Catharine of Braganza, British residents were afforded special favours and privileges, which enabled them to develop the wine trade. Dr. Azevado says that a document exists in the municipal archives of Funchal showing that during the negotiations for the royal marriage, there being some delay in the final decision of King Charles, the Queen Regent of Portugal was willing to cede the island of Madeira as part of her daughter's dowry. Other more important possessions having been ceded, Madeira remained a Portuguese colony, and only came under the protection of the English when, in 1801, in order to protect their allies from the aggressions of the French, the island was garrisoned by English troops. The Peace of Amiens saw the withdrawal of the British forces; but when war broke out between England and France, in 1807, Madeira again came under British protection, when Admiral Hood occupied the island with a force of 4,000 men. Mr. Yate Johnson, in his "Handbook on Madeira," tells us how he himself had seen the original signatures of the principal inhabitants taken on this occasion, by which they individually swore "to bear true allegiance and fealty to His Majesty King George III. and to his heirs and successors, as the island should be held by his said Majesty or his heirs, in conformity to the terms of the capitulation made and signed on the 26th December, 1807, whereby the island and dependencies were delivered over to his said Majesty." The island, though garrisoned by the English until the restoration of general peace in 1814, was restored to her rightful owners four months after the above oath of allegiance was signed.

The year 1826 was a troublous time for Madeira, as the island did not escape the civil war which raged in Portugal in consequence of the Miguelite insurrection. Property was confiscated, the owners being thankful if they escaped with their lives; and even after the country had resumed the monarchy, it took some years before the island returned to its former tranquillity and prosperity.